BAYESIAN EPISTEMOLOGY

Bayesian Epistemology

Luc Bovens
and
Stephan Hartmann

CLARENDON PRESS · OXFORD

OXFORD
UNIVERSITY PRESS

Great Clarendon Street, Oxford OX2 6DP

Oxford University Press is a department of the University of Oxford.
It furthers the University's objective of excellence in research, scholarship,
and education by publishing worldwide in

Oxford New York

Auckland Bangkok Buenos Aires Cape Town Chennai
Dar es Salaam Delhi Hong Kong Istanbul Karachi Kolkata
Kuala Lumpur Madrid Melbourne Mexico City Mumbai Nairobi
São Paulo Shanghai Taipei Tokyo Toronto

Oxford is a registered trade mark of Oxford University Press
in the UK and in certain other countries

Published in the United States
by Oxford University Press Inc., New York

First published 2003

British Library Cataloguing in Publication Data

Data available

Library of Congress Cataloguing in Publication Data

Data available

ISBN 0–19–926975–0

ISBN 0–19–927040–6 (pbk.)

1 3 5 7 9 10 8 6 4 2

Typeset 10.25 on 12.5pt DanteMT
by Kolam Information Services Pvt. Ltd, Pondicherry, India
Printed in Great Britain
on acid-free paper by
Biddles Ltd., King's Lynn, Norfolk

Preface

Bayes is all the rage in philosophy. Metaphysicians discuss the nature of probability, philosophers of religion recast the problem of evil and the argument from design in probabilistic terms, ethicists appeal to decision- and game-theoretic arguments. We ride this wave and examine what probabilistic models have to offer for certain topics in epistemology and for epistemological questions in philosophy of science. Our approach is an engineering approach rather than a foundational approach. Just as engineers do not bother with the foundations of geometry when constructing a bridge, we are consumers of probability theory and the theory of Bayesian Networks and construct models to resolve philosophical questions.

We set this methodology to work on the topics of information, coherence, reliability, confirmation, and testimony. There is a longstanding question in epistemology about how to construct a measure that yields a coherence ordering over sets of propositions and there are various proposals in the literature. In Chapter 1, we present an impossibility result to the effect that there cannot exist such a measure. This has implications for the coherence theory of justification. In Chapter 2, we show how to construct a quasi-ordering that respects the claim that the more coherent a set of propositions is, the greater our degree of confidence ought to be in its content, ceteris paribus. We apply this result to the problem of scientific theory choice. In Chapter 3, we introduce different interpretations of witness reliability into our models and apply them to Condorcet-style jury voting and Tversky and Kahneman's Linda puzzle. In Chapter 4, we apply our models to the confirmation of scientific hypotheses by means of partially reliable test instruments. We show that the variety-of-evidence thesis is false under certain plausible interpretations and assess the Duhem–Quine thesis for positively relevant *versus* independent hypotheses and auxiliaries. In Chapter 5, we address 'too-odd-not-to-be-true' reasoning in the assessment of testimony. This is the curious phenomenon that an initially less plausible report from multiple independent witnesses may elicit more confidence than an initially more plausible report. In the Epilogue, we present some general reflections on the role and the challenges of probabilistic modelling in philosophy.

Our research was supported by the National Science Foundation (Science and Technology Studies, SES 00-8058) and by the Alexander von Humboldt Foundation, the Federal Ministry of Education and Research, and the Program for the Investment in the Future (ZIP) of the German government. Luc Bovens was an Alexander von Humboldt fellow at the University of Konstanz in 1997–8 and a Sofja Kovalevskaja Awardee starting in 2002. We are grateful to André Fuhrmann and Jürgen Mittelstrass for hosting these respective programmes. Stephan Hartmann's research was supported by the TransCoop Program and the Feodor Lynen Program of the Alexander von Humboldt Foundation. Luc Bovens thanks the Department of Philosophy at the University of Colorado at Boulder and Stephan Hartmann thanks the Center for Philosophy of Science at the University of Pittsburgh for providing stimulating work environments. For the same reason, we both thank the Department of Philosophy—and especially Jürgen Mittelstrass and Wolfgang Spohn—and the research group 'Philosophy, Probability and Modelling' at the Center for Junior Research Fellows at the University of Konstanz. Many people have made valuable contributions. We thank Jason McKenzie Alexander, Horacio Arló Costa, Franz Dietrich, Brandon Fitelson, Rolf Haenni, Colin Howson, Stephen Leeds, Patrick Maher, Peter Milne, Graham Oddie, Erik J. Olsson, Gabriella Pigozzi, Tomoji Shogenji, and Barbara Tennis. In particular, we thank Peter Momtchiloff for his encouragement and James Hawthorne, Christian List, Josh Snyder, and Paul Thorn for their detailed comments on the complete manuscript.

L. B.
S. H.

Acknowledgements

We acknowledge permission to use parts of previously published materials in this volume. Chapters 1 and 2 overlap with our 'A Probabilistic Theory of the Coherence of an Information Set' in A. Beckermann and C. Nimtz (*eds.*), *Argument and Analysis* (2001). Chapter 2 overlaps with our 'Solving the Riddle of Coherence', *Mind*, 112 (2003). Chapters 3 and 4 overlap with our 'Bayesian Networks and the Problem of Unreliable Instruments', *Philosophy of Science*, 69 (2002). Chapter 5 overlaps with our 'Too Odd (Not) to Be True: A Reply to Olsson', *British Journal of Philosophy of Science*, 53 (2002) (co-authored with Branden Fitelson and Josh Snyder).

Contents

Introduction

Sarah and Abraham laughed when they were told that Sarah would bear a son, considering that she was 90 years old and he was 100 years old. Did they have good reason to laugh? It is curious that Abraham laughs, since it is God himself who is telling him and God has clearly revealed his identity to Abraham. Sarah is not so fortunate. God appears in the form of three strangers who pay a visit to Abraham in the desert. One of these strangers reports to him that Sarah will bear a son. Sarah overhears the stranger, whose true identity is not revealed to her. It seems that she has good reason to laugh, since it would indeed be out of the ordinary that she and Abraham would still have a child considering their old age. Why are Abraham and Sarah in a different situation? They are both provided with implausible information, but Abraham has reason to believe that the source of the information is reliable, whereas Sarah has no such reason.

Every day we receive information from various sources. Should we believe this information? We ask ourselves how plausible the information is, i.e. how expected it is given our background beliefs, and we ask ourselves how reliable we take the source to be. Certainly we are willing to accept plausible information from a reliable source. But when the information is less plausible, we assess the reliability of the source more carefully. Some information is such that we would want a more reliable source before we would be willing to believe it. To determine whether we are willing to believe new information, we weigh the reliability of the source against the plausibility of the information. This is why Abraham should not have laughed. The information was provided to him by a source that he should have known to be fully reliable. But Sarah is in a different situation. Why should she accept information that is highly implausible from a seemingly unreliable source? It is certainly curious that it is Sarah who is reprimanded for laughing whereas Abraham gets off scot-free.

But now suppose that Abraham tells Sarah that God appeared to him and provided the same information as the stranger. Suppose that Sarah starts waking up feeling nauseous every morning. Like the original report from the stranger, these pieces of information, taken individually, may be insufficient for her to believe that she will bear a child. After all, she may have good reason not to trust Abraham's stories about supernatural occurrences, and nausea is a symptom of many ailments. But the fact that all these pieces of information *cohere*, i.e. that they fit together so well, may well induce her to believe that she will indeed bear a child.

On the other hand, if we receive information that does not fit together well, then we are hesitant to believe it. This is the predicament that Isaac is in. When Isaac is an old man and has lost his vision, he asks his elder son Esau to fetch and prepare him some venison, and he promises that he will bless him. God has told Isaac's wife Rebecca that Esau would serve his younger twin brother Jacob. Hence, Rebecca favours Jacob and urges him to pretend that he is Esau so that he and not Esau will receive their father's blessing. Jacob expresses his concern that this sham will never work since he is less hairy than Esau and his father will recognize him. Rebecca then clothes Jacob in Esau's clothes and covers his hands and neck with goatskin. When Isaac receives Jacob, he is faced with incoherent information. He remarks that the voice is the voice of Jacob, but the hands are the hands of Esau. To form an opinion about who is standing in front of him, he asks his son to come closer. It is the smell of Esau's clothes that leads him to conclude incorrectly that it is Esau. The information provided by his hearing and his touch is too incoherent to form any beliefs about the identity of the person before him. It is only when he acquires an item of information from his sense of smell that coheres with a previous item of information that he is willing to believe that it is Esau whom he is dealing with.

It is not only the plausibility of the information and the apparent reliability of the sources, but also the coherence of the information that may induce us to believe it. However, the coherence of the information is meaningful only if the sources are independent or at least not strongly dependent. Sarah would have been less impressed if she had learned that Abraham had been drawing wishful inferences from her nausea before God allegedly appeared to him, or that the stranger was merely reiterating what he was told earlier by Abraham. The coherence of the information would do very little to her beliefs, and she would be little more confident that she would bear a child than she was on the grounds of her nausea alone. When there is strong dependence between the information sources, coherence does not count for much.

So far we have offered little more than truisms. It is hard to quibble with the claim that we typically believe information that comes from apparently reliable sources, when the information is both plausible and coherent. But this truism lies at the heart of the challenging project of spelling out the precise relationship between these determinants of belief. In Chapter 1, we take on this project within a probabilistic framework. Part of this challenge is to give a precise account of the relevant notion of coherence, so that we can order various information sets according to their relative degree of coherence. This challenge has never been met, despite several valiant attempts. We will show that this challenge cannot be met. There is no measure of coherence that permits the construction of a coherence ordering over information sets. The best one can do is to construct a coherence quasi-ordering, i.e. a relation that is transitive and reflexive, but not necessarily complete. In Chapter 2, we define a measure that induces such a quasi-ordering. With this coherence quasi-ordering in hand, it can then be shown that the more coherent a body of information is, the greater our degree of confidence should be that the information is true, assuming that the reliability of the informers and the plausibility of the information is held constant.

The connection of this project to the coherence theory of justification should make it of interest to epistemologists. The Cartesian sceptic claims that we are never justified in believing any item of information that we acquire about the world through sensory experience, witness reports, etc., because we do not have reason to believe that these sources are reliable. The coherentist responds that the very fact that the set of information items that we acquire through various avenues coheres so well gives us a reason to believe that the information is likely to be true. The project should also be of interest to philosophers of science, for coherence plays a significant role in scientific theory choice. We lend more credence to theories that display a greater degree of coherence, ceteris paribus. Kuhn's notion of *consistency* as a criterion of theory choice may plausibly be interpreted in this vein.

Consider how Sarah's degree of confidence as to whether she will bear a child is affected by her experience of nausea and by the information from the stranger. There is an interesting distinction in how we conceive of the reliability of these information sources. The former source is like a pregnancy test. The test is known to yield some false positives—i.e. cases in which the nausea is not a symptom of pregnancy—and some false negatives—i.e. cases in which the pregnancy is not accompanied by nausea. We may have access to the false-positive and false-negative rates for a given type of test. Now suppose that there are several types of pregnancy tests and that Sarah receives a positive result on each of these tests. With each new test report we may become more confident that Sarah is pregnant. But we would not be inclined to revise our assessment of the reliability of these tests. We generally take the medical information about these tests to be authoritative. On the other hand, when Sarah is told by a series of strangers that she will bear a child, she may laugh off the first one. But as she hears this claim repeated by several independent sources, she may start wondering whether these strangers really do know something that she does not know. After all, it is curious that they provide her with the very same information. So, whereas our assessment of the degree of reliability of medical tests remains fixed, our assessment of the degree of reliability of strangers is variable. Medical tests and strangers are paradigm cases, of course. Sarah may know little about pregnancy tests and she may come to trust them more when they all yield the same results. And she may well be resolved to ignore comments by strangers no matter how many provide her with the same speculations. We distinguish between two ways to think about the reliability of our sources. In some cases, we do not let our assessment of their reliability be affected by the particular reports they generate in our own case. In other cases, our assessment of their reliability will be sensitive to the degree to which there is agreement between these reports. We will adopt each stance as appropriate to the issues we address.

In studying the role of coherence in belief formation in Chapters 1 and 2 we will first assume that the reliability of the sources is fixed. Here reliability is exogenous to the model. We will treat the reliability of the sources as an endogenous variable in the model in Chapter 3. This will permit us to take up the issue of the independence of the sources. As long as we retain the assumption that the sources are genuinely independent, the results in Chapters 1 and 2 remain unaffected. But we will also explore what happens to our assessment of the reliability of the sources and to our degree of confidence that the information is true when we relax the assumption of independence. We apply these models to jury voting in social choice theory and to the Linda puzzle in cognitive psychology.

It seems obvious that the coherence of a jury vote—i.e. the degree of consensus about what the verdict should be—affects our judgement of the reliability of the jurors, which in turn affects our degree of confidence that the majority vote is correct. But this intuition is lost in the framework of the Condorcet Jury Theorem due to the overly strong assumption of juror independence. This leads to the counter-intuitive result that our confidence that the majority vote is correct is a function of the difference between the pro and con votes. Hence, for example, a highly coherent majority vote of ten pro versus zero con votes inspires the same confidence as a highly incoherent majority vote of fifty-five pro versus forty-five con votes. We present a model that is sensitive to the intuition above and avoids this counter-intuitive result.

Tversky and Kahneman (2002: 24) provide the subjects in a psychological experiment with background information about a certain Linda. Given this background information, the subjects find it more plausible that she is a feminist *and* a bank teller than that she is a bank teller *tout court*. Tversky and Kahneman conclude that the subjects are irrational since their judgements violate the Kolmogorov axioms. We show that this need not be the case. The proper comparison is between the posterior probability that Linda is a feminist and a bank teller after being told both that she is a feminist and a bank teller and the posterior probability that she is a bank teller after being told no more than that she is a bank teller. And clearly, the former need not be lower than the latter. We can well imagine that the information source raises the reliability of her reports about Linda by including the information that she is a feminist. Once the information source has a foot in the door, the information that she is a bank teller is passed on without raising too much suspicion. We will construct a model that vindicates the rationality of the experimental subjects in this manner.

In developing our models we have found it helpful to appeal to the theory of Bayesian Networks in Artificial Intelligence. Chapter 3 contains a short introduction to the subject. Although all of our claims can be made without invoking Bayesian Networks, they simplify calculations and make our arguments easier to follow by providing a perspicuous representation of the independence assumptions in our models.

In Chapter 4, we will show how our models are relevant to confirmation theory and, in particular, to scientific testing with partially reliable instruments. We take on two theses in philosophy of science, viz. the variety-of-evidence thesis and the Duhem–Quine thesis. The variety-of-evidence thesis states that more varied evidence provides more confirmation for the hypothesis. We show that this thesis is false under two plausible interpretations of the thesis. We make the assumption that the test instruments are only partially reliable and show, first, that confirming evidence from a single rather than multiple test instruments may provide more confirmation for the hypothesis, ceteris paribus. Second, we show that confirming evidence about a single test consequence rather than about multiple test consequences of the hypothesis may provide more confirmation for the hypothesis, ceteris paribus.

The Duhem–Quine thesis states that when we receive a disconfirming report in testing a hypothesis, the option is always open to reject auxiliary theories rather than the hypothesis itself. We are interested in hypothesis testing when the reliability of the instrument is itself specified by an auxiliary theory of the instrument. In Bayesian treatments of the Duhem–Quine literature it has often been assumed that the hypothesis and the auxiliary theory are independent. This simplifies the calculations, but is not in the spirit of Quinian holism. Scientific practice contains many examples in which the theory of the instrument and the hypothesis are enmeshed in the same web of belief. We will consider how much confirmation a confirming report provides to the hypothesis when the auxiliary theory is positively relevant to the hypothesis as opposed to independent of the hypothesis. It turns out that, under suitable ceteris paribus conditions, less plausible hypotheses receive stronger confirmation when there is a positively relevant auxiliary theory in play, whereas more plausible hypotheses receive stronger confirmation when there is an independent auxiliary theory in play.

In Chapter 5 we impose another twist on modelling the reliability of information sources. Earlier on in *Genesis*, we encounter Noah. After becoming drunk on his first vintage of wine, Noah falls asleep naked in his tent. His son Ham finds him in this condition and tells his brothers Shem and Japheth about it. Shem and Japheth take a cloak to cover their father's naked body. They carefully walk backward so as not to catch a glimpse of their father's unseemly condition. When Noah awakes from his drunken stupor and finds out what happened, he blesses Shem and Japheth and curses poor Ham and his descendants into slavery for the crimes of seeing him naked and telling his brothers about it.

The story says nothing about how Noah learns who the culprit was. Suppose that he gets wind of what happened by listening to his sons but that each one blames the other. So he goes out and asks a few of his friends whether they know anything more. His friends are quite obliging, but considering that they shared in Noah's joys the night before, they are unlikely to be too reliable. Suppose that they all point to Ham. If their reports are independent, this would indeed provide some evidence for Ham being the culprit. But it could also be chance—his friends may not have seen much and they may just happen to pick out poor Ham. However, if, in the style of the times, Noah had ten sons, all considered suspects, then an indictment of Ham would provide stronger evidence, since it would be quite extraordinary that all of Noah's friends would have picked out Ham by chance.

The challenge is to construct a model of what we have termed 'too-odd-not-to-be-true' reasoning. Sometimes we are very much impressed by an odd story told by multiple independent witnesses. The story is so out of the ordinary that it would just be too much of a coincidence if they had each made it up on their own. We expand our earlier model of witness reliability to account for this phenomenon. This discussion should be of special interest to researchers in the area of evidence and probability in jurisprudence.

This book does not aim at providing an overview of the work that has been done in Bayesian epistemology. Rather, we have taken on standard questions in epistemology and have tried to elucidate these questions by appealing to probabilistic models. We conclude in the Epilogue with some general reflections on the virtues of probabilistic modelling in philosophy. Modelling can clarify our intuitions, provide answers to open philosophical questions and be a tool for generating counter-intuitive results. But there are also vices to contend with. Modelling becomes a gratuitous exercise when it does little more than mirror a perfectly good informal argument or leave one to draw conclusions under overly restrictive assumptions. The reader may judge for herself whether the virtues of our models outweigh their vices.

1

Information

1.1. C. I. LEWIS'S HERITAGE

Laurence BonJour (1985: 97, 147–8) draws our attention to some passages in C. I. Lewis's *An Analysis of Knowledge and Valuation*. Lewis argues that the degree of confidence that we have in n information items (for $n \geq 2$) gathered from independent and partially reliable witnesses[1] is positively affected by their *congruence* (1946: 243–53). The core idea is that the more congruent the information is, i.e. the better it meshes or fits together, the more confident we may be that the information is true. On one side of the continuum there is full congruence, viz. when the witnesses all provide us with precisely the same information. On the other side of the continuum there is complete lack of congruence, viz. when the witnesses provide us with items of information that are mutually exclusive. Between these extremes there are various gradations. Suppose that we are informed by one witness that a particular person drives a Porsche and by another witness that he is a millionaire. This is more congruent information than when we are informed by one witness that the person in question drives a Porsche and by another witness that he is homeless. Lewis (1946: 338) only distinguishes between congruent and non-congruent information sets and proposes a probabilistic criterion to draw this distinction.

In BonJour, 'coherence'[2] has been substituted for 'congruence', because of its role in the coherence theory of justification. We will present a precise and seemingly plausible interpretation of Lewis's claim and will name this interpretation 'Bayesian Coherentism'. Then

[1] Lewis actually talks about 'relatively unreliable witnesses' (1946: 346). Following Olsson (2002*b*: 259), we substitute 'partial reliability' for 'relative unreliability'.

[2] Coherence is a property of information *sets*. At some junctions we also talk about the coherence of information, the coherence of reports, or the coherence of particular propositions for ease of presentation. Nothing hangs on this and each such occurrence can readily be rephrased in terms of information sets.

we will construct a model of independent and partially reliable witnesses to evaluate whether Bayesian Coherentism is defensible.

The results of our analysis are twofold. First, on our interpretation of Lewis, Bayesian Coherentism will turn out to be too strong a thesis. Our analysis will show why the quest for a probabilistic measure that induces a coherence ordering over information sets is in vain. Second, our analysis will suggest a way to salvage certain intuitions that underlie Bayesian Coherentism. This can be achieved if we accept that there cannot exist a coherence *ordering*, but only a coherence *quasi-ordering*, over information sets.

1.2. BAYESIAN COHERENTISM

Suppose that we receive items of information from independent and partially reliable sources, say, observations, witness consultations, experimental tests, etc. Then what determines our degree of confidence that the conjunction of these items of information is true? Consider the following procedure. We are trying to determine the *locus* of the faulty gene on the human genome that is responsible for a particular disease. Before conducting the experiments, there are certain *loci* that we consider to be more likely candidates. We run two tests with independent and partially reliable instruments. Each test identifies an area on the human genome where the faulty gene might be located. It turns out that there is a certain overlap between the indicated areas. It is plausible that the following three factors affect our degree of confidence that *both* tests are providing us with correct data, i.e. that the faulty gene is indeed located somewhere in the overlapping area.

(i) *How expected are the results*? Compare two cases of the above procedure. Suppose that the only difference between the cases is that, given our background knowledge, in one case the overlapping area is initially considered to be a highly expected candidate area, whereas in the other case the overlapping area is initially considered to be a highly unexpected candidate area for the faulty gene. Then clearly, our degree of confidence that the *locus* of the faulty gene is in the overlapping area will be lower in the latter than in the former case.

(ii) *How reliable are the tests?* Again, compare two cases of the above procedure. Suppose that the only difference between the cases is that in one case the tests are highly reliable, whereas in the other case they are highly unreliable. Then clearly, our degree of confidence that the *locus* of the faulty gene is in the overlapping area will be lower in the latter than in the former case.

(iii) *How coherent is the information?* This time suppose that the only difference is that in one case both tests identify precisely the same relatively narrow area, whereas in the other case each test identifies a broad area for possible *locus* with an overlap between these areas that coincides with the relatively narrow area in the first case. Then clearly, our degree of confidence that the *locus* of the faulty gene is in the overlapping area will be lower in the latter than in the former case.

The standard way of expressing these claims is as ceteris paribus claims. When gathering information from independent and partially reliable sources, the following claims seem to hold true. First, the more expected (or equivalently, the less surprising) the information is, the greater our degree of confidence, ceteris paribus. Second, the more reliable the information sources are, the greater our degree of confidence, ceteris paribus. Third, the more coherent the information is, the greater our degree of confidence that the information is true, ceteris paribus.

The third claim is a core claim of Bayesian Coherentism. To make it more precise we introduce the following terminology. Let us assume that we obtain the information items $R_1, \ldots, R_n$ from n independent and partially reliable sources. Then $S = \{R_1, \ldots, R_n\}$ is an information set. Now let $\boldsymbol{S}$ be a set of such information sets. Then the following is the first tenet of Bayesian Coherentism:

(BC_1) For all information sets $S, S' \in \boldsymbol{S}$, if S is no less coherent than S', then our degree of confidence that the content of S (i.e. the conjunction of the propositions in S) is true is no less than our degree of confidence that the content of S' is true, ceteris paribus.

What the ceteris paribus clause in (BC_1) indicates is the following. The impact of the coherence of the information set on our degree of confidence satisfies (BC_1), assuming that how expected the information is and how reliable the sources are does not vary from information set to information set. Certainly our degree of confidence in less coherent information is, on occasion, greater than our degree of confidence in more coherent information. This may happen when the less coherent information is more expected or when the corresponding witnesses are more reliable. But a Bayesian Coherentist contends that the coherence of an information set increases our degree of confidence assuming that we keep all other relevant factors, viz. the expectance and the reliability, fixed.

What exactly constitutes the coherence of an information set? A number of proposals have been put forward over the years. Lewis (1946: 338) proposes the following criterion: '*A set of statements, or a set of supposed facts asserted, will be said to be congruent if and only if they are so related that the antecedent probability of any one of them will be increased if the remainder of the set can be assumed as given premises*' (italics in original). Tomoji Shogenji (1999) defends a coherence measure that equals the ratio of the joint probability of all the propositions in S over the product of the marginal probabilities of the propositions in S. Erik J. Olsson (2002*b*: 250) suggests as a possible measure of coherence the ratio of the joint probability of the propositions in S over the probability of the disjunction of the propositions in S. Branden Fitelson (2003) defends a measure of coherence that is based on the Kemeny and Oppenheim measure of factual support. We will return to these proposals in Section 2.6.

There is one thing that all these proposals share, viz. their probabilistic nature. We return to the information set $S = \{R_1, \ldots, R_n\}$ contained in the set of information sets $\mathbf{S}$. Let R_i be the binary propositional variable whose positive value is $\mathrm{R_i}$ and whose negative value is $\neg \mathrm{R_i}$ for $i = 1, \ldots, n$. The *probabilistic features* of an information set are fully expressed by the joint probability distribution over $R_1, \ldots, R_n$. We can express the second tenet that defines Bayesian Coherentism as follows:

(BC_2) A coherence ordering over $\mathbf{S}$ is fully determined by the probabilistic features of the information sets contained in $\mathbf{S}$.

So, for any two information sets $S = \{R_1, \ldots, R_m\}$ and $S' = \{R'_1, \ldots, R'_n\}$, how the coherence of S compares with the coherence of S′ is fully determined by the joint probability distribution over $R_1, \ldots, R_m$ and the joint probability distribution over $R'_1, \ldots, R'_n$. Hence, a Bayesian account of coherence should be able to offer a coherence measure m as a function of the probabilistic features of S and S′ so that S is no less coherent than S′ if and only if $m(S) \geq m(S')$.

For instance, suppose that we are trying to identify the culprit in a murder case. Consider the information set S = {R_1 = [The culprit is French][3], R_2 = [The culprit drove away from the crime scene in a Renault]}. R_1 is the binary propositional variable whose values are R_1 and $\neg R_1$. We assume that, in a population of suspects who stand an equal chance of being culprits, the French are in a minority, but most of the French drive Renaults and Renaults are rarely driven by anyone who is not French. Then we may well have the following joint probabilities: $P(R_1, R_2) = .10, P(R_1, \neg R_2) = .01, P(\neg R_1, R_2) = .01$ and $P(\neg R_1, \neg R_2) = .88$. Intuitively, this information set is highly coherent. Suppose on the other hand that we are dealing with the information set S′ = {R'_1 = [The culprit is French], R'_2 = [The culprit is a Presbyterian]}. We assume that the French and the Presbyterians are both minorities in our population of suspects and that French Presbyterians are very rare indeed. Then the following joint probabilities may hold: $P(R'_1, R'_2) = .01$, $P(R'_1, \neg R'_2) = .10, P(\neg R'_1, R'_2) = .10, P(\neg R'_1, \neg R'_2) = .79$. Intuitively, this information set is strongly incoherent. A measure of coherence should determine whether S ranks higher in the coherence ordering than S′. As an illustration we calculate Shogenji's measure m_s for both information sets. Remember that m_s is the ratio of the joint probability of the propositions in the information set over the product of their marginal probabilities. Thus, in our example:

$$(1.1) \qquad m_s(S) = \frac{P(R_1, R_2)}{P(R_1)P(R_2)} = \frac{.10}{(.10 + .01)(.10 + .01)} \approx 8.26,$$

$$m_s(S') = \frac{P(R'_1, R'_2)}{P(R'_1)P'(R'_2)} = \frac{.01}{(.01 + .10)(.01 + .10)} \approx .826.$$

Since $m_s(S) > m_s(S')$, the Shogenji measure squares with our intuitive ranking of coherence in this case.

We take (BC_1) and (BC_2) not only to be the core of Bayesian Coherentism, but also to have a certain independent plausibility. However,

[3] Following Quine (1960: 168), we use square brackets to refer to the proposition expressed by the enclosed sentence.

we will present an impossibility result to the effect that these theses cannot jointly be true.

1.3. MODELLING INFORMATION GATHERING

Suppose that there are n independent and partially reliable sources and that each source i informs us of a proposition R_i, for $i = 1, \ldots, n$, so that the information set is $\{R_1, \ldots, R_n\}$. Let us call R_i a *fact variable* and $REPR_i$ a *report variable*. $REPR_i$ can take on two values, viz. $REPR_i$ and $\neg REPR_i$. $REPR_i$ is the proposition that, after consultation with the proper source, there is a report to the effect that R_i is the case. $\neg REPR_i$ is the proposition that, after consultation with the proper source, there is no report to the effect that R_i is the case. We construct a probability measure $\boldsymbol{P}$ over $R_1, \ldots, R_n, REPR_1, \ldots,$ and $REPR_n$, satisfying the constraint that the sources are partially reliable and independent.[4]

For the coherence of the reports to be of any consequence, the witnesses must be partially reliable. The chance that the reports of fully reliable witnesses are false is nil, i.e. $\boldsymbol{P}(REPR_i|\neg R_i) = 0$ for $i = 1, \ldots, n$. Our degree of confidence is raised to certainty in whatever fully reliable witnesses report, regardless of the degree of coherence of these reports. On the other hand, fully unreliable witnesses pay no attention whatsoever to the facts on which they are reporting. It is as if they flip a coin or cast a die to determine what they will say. Let the true positive rate be $p_i := \boldsymbol{P}(REPR_i|R_i)$ and let the false positive rate be $q_i := \boldsymbol{P}(REPR_i|\neg R_i)$. Then, for fully unreliable witnesses, $p_i = q_i$ for $i = 1, \ldots, n$. Clearly, the reports of fully unreliable witnesses should be of no consequence to our degree of confidence regarding the matters attested to, regardless of the coherence of the reports.[5] Hence, we stipulate that the witnesses in which we are interested here should be more informative than fully unreliable witnesses yet less informative than fully reliable witnesses,

[4] Our model of partially reliable sources matches interpretation (ii) of 'dubious information-gathering processes' in Bovens and Olsson (2000: 698). Our model of independent sources can be found in Bovens and Olsson (2000: 690 and 696–70 and 2002: 143–4) and in Earman (2000: 56–9).

[5] The reader might ask why fully unreliable witnesses are not modelled as consistent liars, i.e. $\boldsymbol{P}(REPR_i|R_i) = 0$ and $\boldsymbol{P}(REPR_i|\neg R_i) = 1$. The information of consistent liars is actually a very reliable guide to belief formation. We simply need to turn around the truth-value of the report to get to the truth of the matter.

i.e. $p_i > q_i > 0$ for $i = 1, \ldots, n$. To keep things simple, let us assume that all witnesses are equally reliable, i.e. $p_i = p$ and $q_i = q$ for $i = 1, \ldots, n$.[6] Hence, to model partially reliable witnesses we impose the following constraint on ***P***:

(1.2) $$p > q > 0.$$

There are two aspects to the independence of the sources to consider. First, the coherence of the reports is of little consequence when the witnesses have based their reports on information that was communicated between themselves or when they have inferred their reports from facts other than those that they are reporting on. Consider a variation on our earlier example. One witness informs us that the culprit had a French accent and the other witness informs us that the culprit drove off in a Renault. Supposing that most French have French accents and drive Renaults and few non-French have French accents or drive Renaults, the coherence of the witness reports provides a strong boost to our degree of confidence that the culprit is a Renault driver with a French accent. It would indeed be a remarkable coincidence to receive independent witness reports that fit together so well. But the coherence of these reports would be of little consequence if one witness had told the other witness that the culprit drove off in a Renault and the latter had inferred from this information that the culprit had a French accent. The coherence of the reports would also be of little consequence if both sources saw the culprit drive off in what they took to be a Renault and one of the witnesses had inferred from this that the culprit had a French accent.[7] Independent witnesses are supposed to gather information by, and only by, observing the facts they report on. They may not always provide a correct assessment of these facts, but they are not supposed to be influenced by the reports of the other witnesses, nor by the facts on which other witnesses report.

[6] We will show that Bayesian Coherentism is false even in the simple case in which the witnesses are equally reliable. So there is little point in investigating the more complex case involving unequal reliability levels for the various witnesses.

[7] Of course if the witnesses independently observed that the witness drove off in what they took to be a Renault, then the coherence of their observations would increase our confidence that the culprit is a Renault driver. But the coherence of their reports as such would not increase our confidence that the culprit is a Renault driver with a French accent.

We provide the following probabilistic interpretation of what constitutes independent witnesses. Let there be a certain chance p that we will receive a report to the effect that the culprit has a French accent given that he does indeed have a French accent and a certain chance q that we will receive a report to the effect that the culprit has a French accent given that he does not have a French accent. p and q reflect how skilful the witness is at recognizing French accents. Now suppose that we come to learn that the culprit was driving a Renault or we come to learn that another witness reported that the culprit was driving a Renault. Since the independent witness who reported on the culprit's accent strictly attended to the culprit's accent, not to the culprit's car or reports about the car, p and q remain unaffected. This stipulation translates into the following constraint on $\boldsymbol{P}$. All report variables $REPR_i$ are probabilistically independent of all the other fact variables R_j and all the other report variables $REPR_j$, given the fact variable R_i for $i = 1, \ldots, n$. In formal notation:

$$(1.3) \qquad REPR_i \perp\!\!\!\perp R_1, REPR_1, \ldots, R_{i-1}, REPR_{i-1}, R_{i+1}, REPR_{i+1}, \ldots, R_n, REPR_n | R_i, \text{ for } i = 1, \ldots, n.$$ [8]

Or equivalently, in the terminology of the theory of probabilistic causality, we say that R_i *screens off* $REPR_i$ from all other fact variables R_j and from all other report variables $REPR_j$.[9]

The degree of confidence in the information set is the conditional joint probability of the propositions in the information set, given that all the reports have come in, i.e. $\boldsymbol{P}(\mathrm{R}_1, \ldots, \mathrm{R}_n | \mathrm{REPR}_1, \ldots, \mathrm{REPR}_n)$. For simplicity, we suppress the conditionalization by introducing the posterior probability function $\boldsymbol{P}^*$:

$$(1.4) \qquad \boldsymbol{P}^*(\mathrm{R}_1, \ldots, \mathrm{R}_n) = \boldsymbol{P}(\mathrm{R}_1, \ldots, \mathrm{R}_n | \mathrm{REPR}_1, \ldots, \mathrm{REPR}_n).$$

Some additional notational conventions will permit a simple representation of this posterior probability. First, it will prove useful to define a parameter $r := 1 - q/p$ which characterizes the reliability of the wit-

[8] This notation was introduced by Dawid (1979) and has become standard notation. See Pearl (2000) and Spirtes *et al.* (2000).

[9] See Reichenbach (1956) and Salmon (1998).

ness with respect to the report in question.[10] This reliability parameter is a continuous and strictly decreasing function of the *likelihood ratio*[11] q/p—i.e. the proportion of false positives over true positives. The greater this ratio is, the less reliable the witness report is. For fully unreliable witnesses, the false positive rate q equals the true positive rate p and r takes on the value 0, whereas for fully reliable witnesses the false positive rate q equals 0 and r takes on the value 1. Since the witnesses we consider are neither fully reliable nor fully unreliable, r ranges over the open interval (0, 1).

Note that r measures the reliability of the witness *with respect to the report in question* and not the reliability of the witness *tout court*. To see this distinction, consider the case in which q equals 0. In this case, r reaches its maximal value 1, no matter what the value of p is. Certainly, a witness who provides fewer rather than more false negatives, as measured by $1 - p$, is a more reliable witness *tout court*. But when q is 0, the reliability of the witness *with respect to the report in question* is not affected by the value of $p > 0$. No matter what the value of p is, we can be fully confident that what the witness says is true, since $q = 0$—i.e. she never provides any false positives. We will use the elliptical expression of witness reliability to stand for the reliability of the witness *with respect to the report in question*, not for the reliability of the witness *tout court*.

Second, we define parameters a_i for $i = 0, \ldots, n$. Remember that a fact variable R_j can take on a positive value $\mathrm{R_j}$ or a negative value $\neg\mathrm{R_j}$ for $j = 1, \ldots, n$. a_i is the sum of the joint probabilities of all combinations of i negative values and $n - i$ positive values of the variables $R_1, \ldots, R_n$. For example, for an information triple containing the propositions $\mathrm{R_1, R_2}$, and $\mathrm{R_3}$, $a_2 = \boldsymbol{P}(\neg\mathrm{R_1}, \neg\mathrm{R_2}, \mathrm{R_3}) + \boldsymbol{P}(\neg\mathrm{R_1}, \mathrm{R_2}, \neg R_3) + \boldsymbol{P}(\mathrm{R_1}, \neg\mathrm{R_2}, \neg\mathrm{R_3})$. That is, a_2 is the sum of the joint probabilities of all combinations with two negative values and one positive value. Call $<a_0, \ldots, a_n>$ the *weight vector* of the information set $\mathrm{S} = \{\mathrm{R_1}, \ldots, \mathrm{R_n}\}$ and note that $\sum_{i=0}^{n} a_i = 1$.

We show in Appendix A.1 that, given the constraints on $\boldsymbol{P}$ in (1.2) and (1.3), the following relationship holds:

[10] We will later show that the results in Chapters 1 and 2 do not depend on this particular choice of a reliability measure.

[11] One needs to be careful when talking about the likelihood ratio in Bayesian confirmation theory. Sometimes the likelihood ratio is defined as above (e.g. in Howson and Urbach 1993: 29), sometimes as the reciprocal p/q (e.g. in Pearl 1997: 34).

$$P^*(R_1, \ldots, R_n) = \frac{a_0}{\sum_{i=0}^{n} a_i \bar{r}^i}. \tag{1.5}$$

in which $\bar{r} := 1 - r = q/p$. For a better understanding of this formula, consider the following two situations. In the first situation we choose three propositions that are the positive values of independent propositional variables. The witnesses tell us, respectively, that the culprit (R_1) was a woman, (R_2) had a Danish accent, and (R_3) drove a Ford. Suppose that our population of suspects is composed so that learning that one or two of these propositions are true (or false) does not change the probability of the other proposition(s). The diagram in Figure 1.1 depicts a possible joint probability distribution over the variables R_1, R_2, and R_3 and presents the corresponding values for a_i, for $i = 0, \ldots, 3$. In the second situation we choose three equivalent propositions. The witnesses tell us, respectively, that the culprit (R'_1) was wearing Coco Chanel shoes, (R'_2) had a French accent, and (R'_3) drove a Renault. Our population of suspects is composed so that all and only people with French accents wear Coco Chanel shoes, and all and only people who wear Coco Chanel shoes drive Renaults. The diagram in Figure 1.2 depicts a possible joint probability distribution over the variables R'_1, R'_2, and R'_3 and presents the corresponding values for a'_i, for $i = 0, \ldots, 3$. Note that $a_0 = a'_0$: The prior joint probability of the information in the former information set is equal to the prior

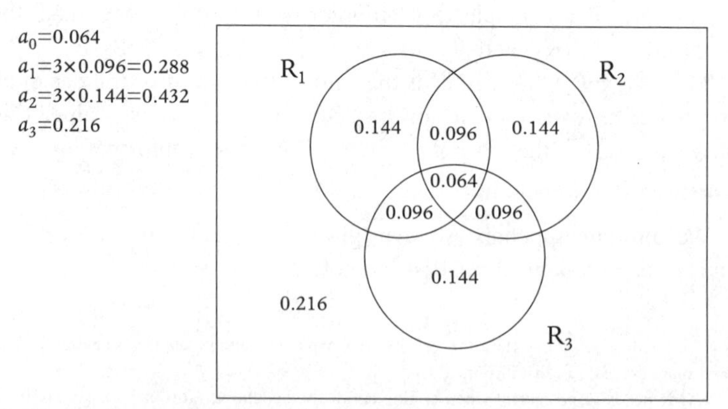

Fig. 1.1 A diagram of the joint probability distribution over the variables R_1, R_2, and R_3

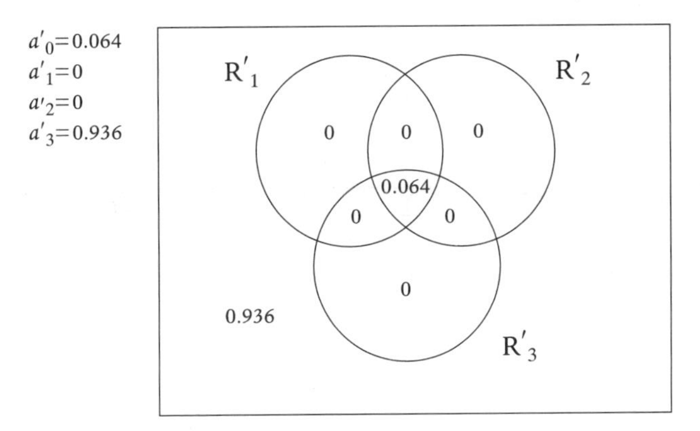

FIG. 1.2 A diagram of the joint probability distribution over the variables R'_1, R'_2, and R'_3

joint probability of the information in the latter information set. In other words, the information is equally expected in both situations. Suppose that the sources are twice as likely to issue a true positive report than a false positive report, i.e. $p = 2q$ and hence $r = \bar{r} = .50$. Then, by (1.5), our degrees of confidence after we have received the reports from the sources in these two situations are, respectively,

$$(1.6)\quad P^*(\mathrm{R}_1, \mathrm{R}_2, \mathrm{R}_3) = \frac{.064}{.064 \times .50^0 + .288 \times .50^1 + .432 \times .50^2 + .216 \times .50^3} \approx .187 \text{ and}$$

$$(1.7)\quad P^*\left(\mathrm{R}'_1, \mathrm{R}'_2, \mathrm{R}'_3\right) = \frac{.064}{.064 \times .50^0 + 0 \times .50^1 + 0 \times .50^2 + .936 \times .50^3} \approx .354.$$

Notice that for equally expected information and equally reliable sources, the posterior probability is greater in the second situation than in the first. And indeed, the information does fit together more tightly in the second situation. Hence, the comparison of these situations constitutes one example of the impact of relative coherence that is fully consistent with the tenets of Bayesian Coherentism. But we will now show that Bayesian Coherentism does not hold in general.

1.4. AN IMPOSSIBILITY RESULT

To disprove Bayesian Coherentism, it will suffice to construct a single counter-example to (BC_1) and (BC_2). Note that for information triples,

$a_0 + a_1 + a_2 + a_3 = 1$. Hence, from (1.5), our degree of confidence that the reports from three independent information sources are true is given by the formula:

$$P^*(\mathrm{R}_1, \mathrm{R}_2, \mathrm{R}_3) = \frac{a_0}{a_0 + a_1\bar{r} + a_2\bar{r}^2 + (1 - a_0 - a_1 - a_2)\bar{r}^3}. \tag{1.8}$$

Now pick any two information triples S and S′ with joint probability distributions that yield the respective weight vectors $< a_0, a_1, a_2, a_3 >=$ $< .05, .30, .10, .55 >$ and $< a'_0, a'_1, a'_2, a'_3 >= < .05, .20, .70, .05 >$. The posterior joint probability that the information is true is plotted in Figure 1.3.

We do not know how to construct a coherence measure for information triples. But this does not matter. Our strategy will be to show that *any* coherence measure would leave (BC_1) and (BC_2) vulnerable to counter-examples. Hence, no reasonable proposal for a coherence measure could ever succeed.

By (BC_2), a coherence measure that induces an ordering should be a function of the probabilistic features of the information set. Since the weight vector is the only relevant information of the probability distribution in determining our degree of confidence that the information is

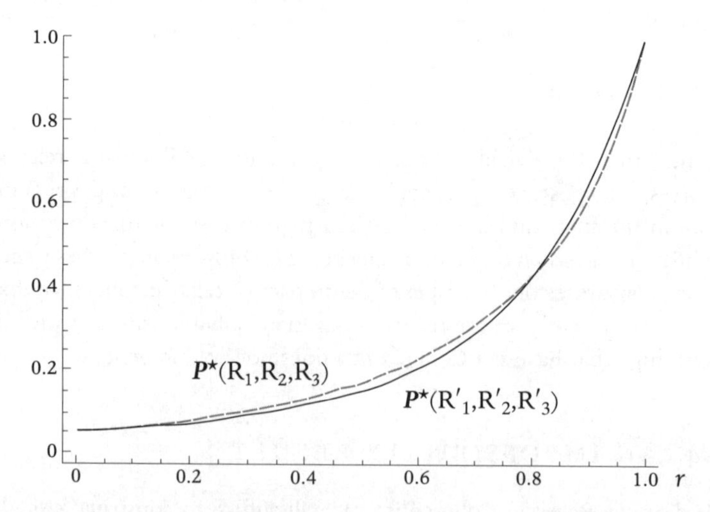

FIG. 1.3 The posterior probability for information triples with weight vectors $< a_0, a_1, a_2, a_3 > = < .05, .3, .1, .55 >$ and $< a'_0, a'_1, a'_2, a'_3 > = < .05, .2, .7, .05 >$ as a function of the reliability parameter r

true, the measure should be a function of only $<a_0, a_1, a_2, a_3>$ and $<a'_0, a'_1, a'_2, a'_3>$. However, for the present example, whatever measure we choose will violate (BC_1). To see this, first notice that the information in S and S′ is equally expected, since $a_0 = a'_0$. Suppose that we pick a measure $m_\dagger$ so that $m_\dagger(S') \geq m_\dagger(S)$. Then for any value of $r \in (.8, 1)$, (BC_1) is violated. It is not true that the more coherent the information is, the greater our degree of confidence, ceteris paribus, since $P^*(R'_1, R'_2, R'_3) < P^*(R_1, R_2, R_3)$ over this interval. Or suppose that we pick a measure $m_\ddagger$ so that $m_\ddagger(S) > m_\ddagger(S')$. Then for any value of $r \in (0, .8]$, (BC_1) is false. It is not true that the more coherent the information is, the greater our degree of confidence, ceteris paribus, since $P^*(R_1, R_2, R_3) \leq P^*(R'_1, R'_2, R'_3)$ over this interval. Thus, no measure of coherence can be constructed that determines our relative degree of confidence when all other determinants, i.e. the expectance of the information and the reliability of the witnesses, remain the same for both information sets. For the weight vectors in question, the reliability of the sources changes which information set will merit the greater degree of confidence. Similar results can be generated for information sets of size $n > 3$.[12] Hence, we can conclude that there cannot exist a measure of coherence that is probabilistic and induces a coherence ordering for information triples (BC_2) and that simultaneously makes it the case that the more coherent the information set, the more confident we are that the information is true, ceteris paribus (BC_1).

One might raise the following objections. First, we have shown that our degree of confidence is a function of the reliability r and the weight vector $<a_0, \ldots, a_n>$. It may well be the case that there is another determinant D of our degree of confidence which differs from reliability, expectance, and coherence and which is also a function of r and $< a_0, \ldots, a_n >$. (BC_1) may well be true if we keep the reliability, the expectance, *as well as D* fixed under the ceteris paribus clause. We do not have a general argument to the effect that there is no such determinant D. However, to successfully revive Bayesian Coherentism it will have to be the case that in our counter-example D has no common value in the region $r = (0, .8]$ and in the region $r = (.8, 1)$. If there are any two points in these respective regions for which D is the

[12] It is not possible to construct counter-examples of this nature for information pairs, i.e. information sets of size $n = 2$. This does not mean that Bayesian Coherentism is true for information pairs. In Chapter 2, we will show that there also does not exist a coherence ordering over the set of information pairs, violating (BC_2).

same, our counter-example will continue to apply. We cannot begin to respond to this objection without a hint of what such a determinant D might be.[13]

Second, one might object that our result is an artefact of the specific choice of the reliability measure r. However, our result holds for any measure in the class of measures that are a continuous and a strictly monotonically decreasing function of the likelihood ratio $x = q/p$. This is so because the curves of the posterior probability functions that criss-cross when plotted against r will also criss-cross when plotted against $x = 1 - r$, and hence against any continuous and strictly monotonically decreasing function of x that maps the interval (0, 1) onto the interval (0, 1). Note that the reliability measure only depends on x, and not also on, say, q. To see this, suppose that the measure were to depend on both x and q. We keep x constant and change the value of q (and accordingly the value of p). The witness reliability would thereby change, whereas, by (1.5), the posterior probability of the information would remain constant, which is unintuitive.

1.5. WEAK BAYESIAN COHERENTISM

How troubling should this negative result be? Curiously, our model at first seemed to leave some hope for a probabilistic account of coherence, but then we were able to show that Bayesian Coherentism does not hold up for *certain* pairs of information *triples*. This negative result hinges on a stipulation of the weight vectors of the pairs of information *triples* so that (i) the prior joint probabilities of the propositions in the information triples are the same and (ii) the curves of the posterior joint probability as a function of the reliability of the witnesses criss-cross—i.e. for some values of r the posterior joint probability of the propositions in one information set exceeds the posterior joint probability of the propositions in the other information set, and vice

[13] Furthermore, one might object that, if there exists such a determinant D, then coherence may well be a function of $< a_0, \ldots, a_n >$ *and* some other feature d of the probability distribution so that the coherence measure m is *not exclusively* a function of $< a_0, \ldots, a_n >$. d may be some marginal probability, as in the Shogenji measure. This could be so, as long as D is also a function of d, so that our degree of confidence is independent of d. For example let $P^*(R_1, \ldots, R_n) \sim cD$, with $m = d\, g_1(< a_0, \ldots, a_n >)$ and $D = d^{-1} g_2(< a_0, \ldots, a_n >)$. But once again, what determinant D could qualify for this role?

versa for other values of *r*. But note that this criss-crossing of curves does not occur for *all* pairs of information triples that satisfy condition (i). It will be instructive to compare concrete examples of pairs of information triples in which this criss-crossing occurs with examples of pairs of information triples in which this criss-crossing does not occur.

First, let us consider a case in which this criss-crossing does not occur. We return to the information triples in Section 1.3, viz. $S = \{R_1 =$ [the culprit was a woman], $R_2 =$ [the culprit had a Danish accent], $R_3 =$ [the culprit drove a Ford]$\}$ and $S' = \{R'_1 =$ [the culprit was wearing Coco Chanel shoes], $R'_2 =$ [the culprit had a French accent], $R'_3 =$ [the culprit drove a Renault]$\}$. Suppose that, given background information about the suspects, the weight vectors are respectively $<a_0, \ldots, a_3> = <.064, .288, .432, .216>$ and $<a'_0, \ldots, a'_3> = <.064, 0, 0, .936>$. Figure 1.4 plots the posterior joint probability that the information in each respective triple is true as a function of the reliability measure. Notice that these curves do not criss-cross. Hence, our degree of confidence in the information content of S is greater than our degree of confidence in the information content of S′, no matter at what level we fix the degree of partial reliability of the witnesses. For this

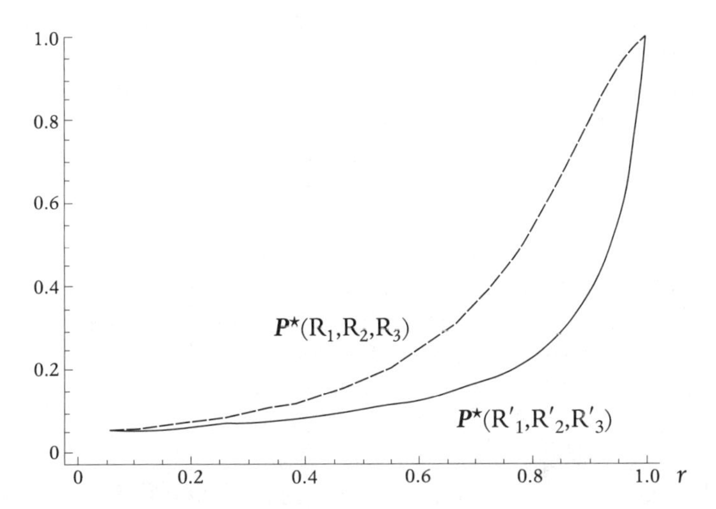

Fig. 1.4 The posterior probability for information triples with weight vectors $<a_0, a_1, a_2, a_3> = <.064, .288, .432, .216>$ and $<a'_0, a'_1, a'_2, a'_3> = <.064, 0, 0, .936>$ as a function of the reliability parameter r

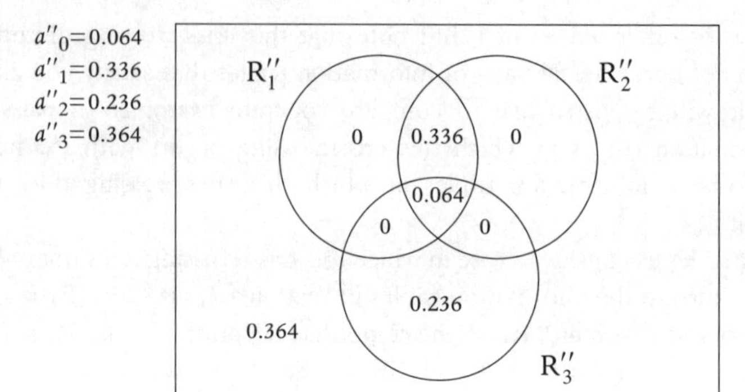

FIG. 1.5 A diagram of the joint probability distribution over the variables R''_1, R''_2, and R''_3

pair of information triples, the tenets of Bayesian Coherentism succeed.

But now compare the following two information triples. The triple S is as before, but the triple $S'' = \{R''_1 =$ [the culprit was wearing Coco Chanel shoes], $R''_2 =$ [the culprit had a French accent], $R''_3 =$ [the culprit drove a Ford]$\}$. Suppose that in our population 40 per cent of suspects wear Coco Chanel shoes, and all and only suspects who wear Coco Chanel shoes have a French accent. Also, 30 per cent of our suspects drive Fords. However, 84 per cent of the suspects who wear Coco Chanel shoes and have a French accent drive Renaults and only the remaining 16 per cent of them drive Fords. We have represented the probability distribution for S'' and calculated the weight vector $<.064, .336, .236, .364>$ in the diagram in Figure 1.5.[14] Which of these two information sets has the appearance of being more coherent? On the one hand, one might say that S is less coherent, since the propositions in S are probabilistically independent whereas S'' has two propositions, viz. R''_1 and R''_2, that are maximally positively relevant, i.e. they pick out coextensive sets of suspects. On the other hand, one might say that S is more coherent, since S'' contains a proposition R''_3

[14] Observe in the diagram that $.336 + .064 = .40$; 40% of the suspects wear Coco Chanel shoes and have a French accent; $.236 + .064 = .30$; 30% of the suspects drive Fords; $.064/.40 = .16$; 16% of the suspects with Coco Chanel shoes and French accents drive Fords. We assume that the remaining 84% of the suspects with Coco Chanel shoes and French accents drive Renaults: $.336/.40 = .84$.

that is highly negatively relevant with respect to the other propositions R''_1 and R''_2. S'' seems more coherent than a set of three independent propositions, since R''_1 and R''_2 fit together so well. Yet S'' also seems less coherent, since R''_3 fits so poorly with R''_1 and R''_2. We want to say that there just is no fact of the matter as to whether S is more or less coherent than S''. No coherence ranking can be defined over the pair $\{S, S''\}$. And indeed, if we plot the posterior joint probability for these two information sets, then we find criss-crossing lines just as in Figure 1.3.[15]

So why is it that the conjunction of (BC_1) and (BC_2) does not generally hold? We can split up (BC_2) into two components:

(BC_2^i) The *binary relation* of '. . . being no less coherent than . . .' over ***S*** is fully determined by the probabilistic features of the information sets contained in ***S***.

(BC_2^{ii}) The *binary relation* of '. . . being no less coherent than . . .' is an *ordering*.

A weakened variant of Bayesian Coherentism can be salvaged if we are willing to give up (BC_2^{ii}). *Orderings* are complete, reflexive, and transitive binary relations; *quasi-orderings* are reflexive, transitive, but not necessarily complete binary relations.[16] Our suggestion is that the Bayesian Coherentist give up on the completeness requirement on the binary relation of coherence. In other words, (BC_2) needs to be replaced by

(BC_2^*) A coherence quasi-ordering over ***S*** is fully determined by the probabilistic features of the information sets contained in ***S***.

Let us name the conjunction of (BC_1) and (BC_2^*) 'Weak Bayesian Coherentism'. According to Weak Bayesian Coherentism, there exists a coherence quasi-ordering over ***S*** that is fully determined by the probabilistic features of its constituent information sets. Furthermore, if S is no less coherent than S', then our degree of confidence that S is

[15] The crossing point can be obtained analytically by solving $P^*(R_1, R_2, R_3) = P^*(R''_1, R''_2, R''_3)$ for $r \in (0,1)$ using equation (1.5). This point lies at $r \approx .68$.

[16] Various terms have been used in the literature. See Sen (1970: 7–9).

true is no less than our degree of confidence that S' is true, ceteris paribus. In our example, an ordering is defined over $\{S, S'\}$ but not over $\{S, S''\}$.

How does our analysis affect the coherence theory of justification? The coherence theory is meant to be a response to Cartesian scepticism. The Cartesian sceptic claims that we are not justified in believing the story about the world that we have come by through various information-gathering processes (our senses, witnesses, etc.), since we have no reason to believe that these processes are reliable. There are many variants of the coherence theory of justification. We are interested in versions that hinge on the claim that it is the very coherence of the story of the world that gives us a reason to believe that the story is likely to be true. This is not the place to defend a full-fledged version of the coherence theory of justification, but we will argue that the substitution of (BC_2^*) for (BC_2) is not damaging to this claim.

First consider the following analogy. Suppose that we establish that the more a person reads, the more cultured she is, ceteris paribus. We conclude from this that if we meet with a very well-read person, then we have a reason to believe that she is cultured. It may not be sufficient reason, but it is a reason nonetheless. Now suppose that we also establish that sometimes no comparison can be made between the amount of reading two people do, since reading comes in many shapes and colours. We can only establish a quasi-ordering over a set of persons according to how well read they are. This does not stand in the way of our conclusion.

We have shown that, as long as our sources are independent and partially reliable, the more coherent an information set is, the more likely its content is to be true, ceteris paribus. We conclude from this that, if the story of the world is a very coherent information set, then we have a reason to believe that its content is likely to be true. Again, it may not be sufficient reason, but it is a reason nonetheless. And similarly, the fact that we can only establish a coherence quasi-ordering over information sets does not stand in the way of this conclusion.

What is misguided in the coherence theory of justification is the persistent attempt to construct a measure that imposes an ordering on sets of information sets. The coherence theory is thought to be lacking unless we have a clear measure of coherence that permits us to order information sets. What we have shown is that the insistence on such a measure is wrong-headed, since there simply is no such measure that also respects (BC_1). A coherence theory that draws on a probabilistic measure of coherence must make do with a quasi-ordering.

Opponents of the coherence theory may try to get some mileage out of our result. Indeed, a radical response to what has been demonstrated would be to discard the Coherentism part of Bayesian Coherentism—i.e. (BC_1). But our formal model actually discourages this move. Our model shows that if S is indeed more coherent than S′, then our degree of confidence in the content of S should be greater than in the content of S′, assuming that the sources are equally reliable and the information is equally expected. The opponent of the Coherentism part of Bayesian Coherentism will need to show that this formal model is not fit to deal with Cartesian scepticism.

A more moderate response to our analysis would be to tinker with (BC_2), i.e. with the Bayesian part of Bayesian Coherentism. We have proposed altering (BC_2^{ii}), but others have tried to tinker with (BC_2^{i}). The idea underlying such proposals is that probabilistic accounts of coherence cannot do justice to the richness of this concept, and that negative results are to be expected when one works within such an impoverished structure. For instance, it has been argued that the coherence of an information set should take into account explanatory relations between the propositions in the set, and that these relations cannot be adequately represented by probabilistic information (e.g. BonJour 1985: 99–101). If one takes into account the full richness of the notion of coherence, then it is possible to construct an ordering—and hence to respect (BC_2^{ii})—of information sets, or so the argument goes. At this point in time we may not have a principled way of doing so, but proponents argue that *this* is the challenge that the coherence theory of justification must take up. We take no dogmatic stand on this issue, but remain suspicious of any claim to the effect that there are aspects of uncertain reasoning that resist a strictly probabilistic analysis.

2

Coherence

2.1. UNEQUAL PRIORS

In the previous chapter we showed that there cannot be a measure that induces a coherence ordering—i.e. a binary relation which is complete, reflexive, and transitive—over the set of possible information sets. This does not exclude the construction of a measure that induces a coherence quasi-ordering—i.e. a binary relation which is reflexive and transitive. So far we have only considered a special case—we have laid out a procedure to order pairs of equal-sized information sets that share the same prior probability that their respective constitutive propositions are all true. In effect, we have partitioned the set of all information sets into subsets **S** of information sets that have the same cardinality and the same prior joint probability a_0. Within each of these subsets **S** we have constructed a procedure to impose a quasi-ordering over **S**. Let '$\succeq$' be the binary relation of *being no less coherent than*. Then for pairs of information sets $S = \{R_1, \ldots, R_n\}$ and $S' = \{R'_1, \ldots, R_n'\}$, our procedure can be stated as follows:

(2.1) For all S, $S' \in \mathbf{S}$, if S and S′ have the same cardinality and $\boldsymbol{P}(R_1, \ldots, R_n) = a_0 = a'_0 = \boldsymbol{P}(R'_1, \ldots, R'_n)$, then $S \succeq S'$ iff $\boldsymbol{P}^*(R_1, \ldots, R_n) \geq \boldsymbol{P}^*(R'_1, \ldots, R'_n)$ for all values of the reliability parameter $r \in (0, 1)$.

In other words, S is no less coherent than S′ if and only if the curve representing the function for their posterior joint probability for S is strictly above the curve for S′ over the interval $r \in (0, 1)$. We have assumed that the witnesses are equally reliable and will discuss this assumption in Section 2.4.

We should be able to do better than this. Our intuitive notion of one information set being no less coherent than another information set is not restricted to information sets whose content is equally probable nor to information sets of the same cardinality. Let us look at a few examples.

First, suppose that a murder has been committed in Tokyo. We are trying to locate the corpse and, given our background knowledge, every square inch of Tokyo is just as likely a spot as every other square inch. Suppose two witnesses independently point to a particular house. This is certainly coherent information. Alternatively, suppose that one witness points to some broad area on the map and the other witness points to an area that is no less broad. The overlap between both areas is a large district of Tokyo. There is little doubt that the information in the first case is more coherent than the information in the second case. And yet the prior probability that the information of the witnesses in the first case is true is much lower than the prior probability that the information of the witnesses in the second case is true, for the house is a much smaller region than the district.

Second, BonJour poses the following example of information sets that can clearly be ordered with respect to their relative coherence. Consider the following two information sets: $S = \{$[All ravens are black], [This bird is a raven], [This bird is black]$\}$ and $S' = \{$[This chair is brown], [Electrons are negatively charged], [Today is Thursday]$\}$ (1985: 96). There is no doubt that set S is more coherent than set S'. And yet there is no reason to assume that the prior probability that the information in S is true equals the prior probability that the information in S' is true.

Third, we also make judgements of relative coherence when the information sets are of unequal size. For instance, consider the paradigm case of non-monotonic reasoning. Certainly the information *pair* $S = \{$[My pet Tweety is a bird], [My pet Tweety cannot fly]$\}$ is less coherent than the information *triple* $S' = \{$[My pet Tweety is a bird], [My pet Tweety cannot fly], [My pet Tweety is a penguin]$\}$. The inclusion of the information that Tweety is a penguin is what brings coherence to the story. What we want is a measure that induces a coherence quasi-ordering over information sets in general, not just information sets of the same size and with equal prior joint probabilities.

Various attempts have been made to provide a probabilistic account of the notion of coherence. In the previous chapter we showed that the search for a measure that imposes a coherence ordering on the set of information sets is in vain. However, a coherence quasi-ordering should suffice for the purposes of the coherence theory of justification. Thus, in this chapter, we will take on the project of showing how to construct a general measure that imposes a coherence quasi-ordering on the set of information sets.

The notion of coherence also plays a role in philosophy of science. Kuhn (1977: 321–2, quoted in Salmon (1990: 176)) mentions *consistency* as one of the (admittedly imprecise) criteria for scientific theory choice (along with accuracy, scope, simplicity, and fruitfulness). Salmon (1990: 198) distinguishes between the internal consistency of a theory and the consistency of a theory with other accepted theories. In discussing the latter type of consistency, he claims that there are two aspects to this notion, viz. the '*deductive* relations of entailment and compatibility' and the '*inductive* relations of fittingness and incongruity'. We propose to think of the internal consistency of a theory in the same way as Salmon thinks of the consistency of a theory with accepted theories. Hence, the *internal consistency* of a theory matches the epistemologist's notion of the *coherence* of an information set: How well do the various components of the theory fit together, how congruous are these components? Salmon also writes that this criterion of consistency 'seem[s] . . . to pertain to assessments of the prior probabilities of the theories' and 'cr[ies] out for a Bayesian interpretation' (1990: 198). Following this line of thought, we will show how one can construct a coherence quasi-ordering over a set of scientific theories and how our relative degree of confidence that one or another scientific theory is true is functionally dependent on this quasi-ordering. That the relation is a quasi-ordering rather than an ordering respects Kuhn's contention that consistency is an imprecise criterion of theory choice. Indeed, in some cases, it is indeterminate which of two theories is more coherent.

2.2. CONSTRUCTING A MEASURE

We will construct a formal measure that permits us to read off a coherence quasi-ordering from the joint probability distributions over the propositional variables whose positive values are constitutive of the information sets. The problem with existing accounts of coherence is that they try to bring precision to our intuitive notion of coherence independently of the particular role that it is meant to play. This is a mistake. To see this, consider the following analogy. We not only use the notion of coherence when we talk about information sets, but also, for example, when we talk about groups of individuals. Group coherence tends to be a good thing. It makes ant colonies more fit for survival, it makes law firms more efficient, it makes for happier families, etc. It makes little sense to ask what makes for a more coherent group independently of the particular role that coherence is supposed to play in the context in question. We must first fix the context in which coherence purports to play a particular role. For instance, let the context be ant colonies and let the role be that of promoting reproductive fitness. We give more precise content to the notion of coherence in this context by letting coherence be the property of ant colonies that plays the role of boosting fitness and at the same time matches our pre-theoretic notion of the coherence of social units. A precise fill-in for the notion of coherence will differ as we consider fitness boosts for ant heaps, efficiency boosts for law firms, or happiness boosts for families.

Similarly, it makes little sense to ask precisely what makes for a more coherent information set independently of the particular role that coherence is supposed to play. The coherence theory of justification and the Kuhnian appeal to coherence as a criterion of theory choice ride on a particular common-sense intuition. When we gather information from independent and partially and equally reliable sources, the more coherent the story is, the more confident we are that the story is true, ceteris paribus. Within the context of information gathering from such sources, coherence is a property of information sets that plays a confidence-boosting role.

In the previous chapter we derived a parsimonious expression for the posterior probability that the information is true which we receive from independent witnesses who are partially and equally reliable:

$$P^*(\mathrm{R}_1, \ldots, \mathrm{R}_n) = \frac{a_0}{\sum_{i=0}^{n} a_i \bar{r}^i}. \tag{2.2}$$

Remember that $\bar{r} := 1 - r$, with r being the reliability parameter equal to $1 - q/p$. The true positive rate $p := P(\mathrm{REPR_i}|\mathrm{R_i})$ is greater than the false positive rate $q := P(\mathrm{REPR_i}|\neg\mathrm{R_i})$ which is greater than 0 for

$i = 1, \ldots, n$. $<a_0, \ldots, a_n>$ is the weight vector of the information set $S = \{R_1, \ldots, R_n\}$. Each a_i is the sum of the joint probabilities of all combinations of i negative values $\neg R_j$ and $n - i$ positive values R_j of the propositional variables $R_1, \ldots, R_n$.

A maximally coherent information set has the weight vector $<a_0, 0, \ldots, 0, \bar{a}_0>$ with $\bar{a}_0 := 1 - a_0$. Let us assume that we are neither certain that the content of the information set is true nor certain that it is false. All items of information $R_1, \ldots, R_n$ are equivalent, since $a_0 = \boldsymbol{P}(R_1, \ldots, R_n)$ and $\bar{a}_0 = a_n = \boldsymbol{P}(\neg R_1, \ldots, \neg R_n)$ and the joint probabilities of all other combinations of propositions are set at 0. If one of the remaining $a_1, \ldots$, or a_{n-1} exceeds 0, then the items of information are no longer equivalent and the information set loses its maximal coherence. It is some feature of $<a_0, \ldots, a_n>$ that determines the coherence of the information set. For maximal coherence, it needs to be the case that $a_i = 0$ for $i = 1, \ldots, n - 1$. But it is not clear at all what feature we are looking for when assessing and comparing cases of non-maximal coherence.

To determine this feature, here is how we will proceed. Suppose that we have a range of suspects for some crime. We question the witnesses, who provide us information about what car the culprit was driving, the culprit's accent, etc. All this information picks out a certain subset of the original suspects that satisfy all these features. Let's suppose that only Jean and Pierre satisfy these features. The information that led us to pick out Jean or Pierre may have been maximally coherent. For instance, it may be the case that each witness provided a report that it was either Jean or Pierre who was the culprit. Or it may be the case that one witness claimed that the culprit is from Marseille and the other witness claimed that the culprit is a sailor and that all and only inhabitants from Marseille are sailors in our population of suspects. But the information may also have been less coherent. One witness might have said that the suspect had a French accent and the other witness that the suspect was a Presbyterian. The population of suspects contains a large subset of suspects with French accents and a large subset of suspects who are Presbyterians, but only Jean and Pierre are Presbyterians with French accents. We learned in the last chapter that for any particular value of the reliability parameter r, our confidence boost that either Jean or Pierre is the suspect is much greater when the information comes to us in the form of maximally coherent information rather than in the form of less than maximally coherent information. Our strategy will be to assess the coherence of an information set by measuring the proportion of the confidence boost that we actually receive, relative to the confidence boost that we would have received *had we received this very same information in the form of maximally coherent information.*

To put this formally, let us turn to our example of independent tests that identify sections on the human genome that may contain the locus of a genetic disease. The tests pick out different areas, and the overlap between the areas is a region σ. The information is more coherent when the reports are all clustered around the region σ than when they are scattered all over the human genome but have this relatively small area of overlap on the region σ. The information is maximally coherent when every single test points to the region σ. We assign a certain prior probability that the locus of the disease is in the region σ. With more coherent reports, our confidence boost will be greater than with less coherent reports. Let us measure this confidence boost by the ratio of the posterior probability—i.e. the probability after we have received the reports—over the prior probability that the locus of the disease is in region σ:

$$b(\{\mathrm{R}_1, \ldots, \mathrm{R}_n\}) = \frac{\boldsymbol{P}^*(\mathrm{R}_1, \ldots, \mathrm{R}_n)}{\boldsymbol{P}(\mathrm{R}_1, \ldots, \mathrm{R}_n)}. \tag{2.3}$$

To determine this confidence boost it is sufficient to know the weight vector $<a_0, \ldots, a_n>$ and the reliability parameter r, since $\boldsymbol{P}(\mathrm{R}_1, \ldots, \mathrm{R}_n)$ equals a_0 and since $\boldsymbol{P}^*(\mathrm{R}_1, \ldots, \mathrm{R}_n)$ is a function of the weight vector and the reliability parameter.

If we had received the information that the locus of the disease is in region σ in the form of maximally coherent information, then our information set would have contained n reports to the effect that the locus of the disease was in region σ, i.e. $\left\{\mathrm{R}_1^\sigma, \ldots, \mathrm{R}_n^\sigma\right\}$. We can impose a probability measure $\boldsymbol{P}^{\max}$ over the propositional variables $R_1^\sigma, \ldots, R_n^\sigma$ with the corresponding weight vector $<a_0, 0, \ldots, 0, a_n>$. We insert this weight vector into (2.2) and calculate what our degree of confidence would have been that the locus of the disease is in region σ, had we received the information as maximally coherent information:

$$\boldsymbol{P}^{\max *}\left(\mathrm{R}_1^\sigma, \ldots, \mathrm{R}_n^\sigma\right) = \frac{a_0}{a_0 + \bar{a}_0 \bar{r}^n}. \tag{2.4}$$

Hence, our confidence boost would have been

$$(2.5) \qquad b^{\max}(\{R_1, \ldots, R_n\}) = \frac{P^{\max*}\left(R_1^\sigma, \ldots, R_n^\sigma\right)}{P^{\max}\left(R_1^\sigma, \ldots, R_n^\sigma\right)}.$$

Since the prior probability $P^{\max}\left(R_1^\sigma, \ldots, R_n^\sigma\right) = P(R_1, \ldots, R_n) = a_0$, the proportion of the confidence boost that we actually receive, relative to the confidence boost that we would have received, had we received this very same information in the form of maximally coherent information, equals

$$\begin{aligned}(2.6) \qquad c_r(\{R_1, \ldots, R_n\}) &= \frac{b(\{R_1, \ldots, R_n\})}{b^{\max}(\{R_1, \ldots, R_n\})} \\ &= \frac{P^*(R_1, \ldots, R_n)/P(R_1, \ldots, R_n)}{P^{\max*}\left(R_1^\sigma, \ldots, R_n^\sigma\right)/P^{\max}\left(R_1^\sigma, \ldots, R_n^\sigma\right)} \\ &= \frac{P^*(R_1, \ldots, R_n)}{P^{\max*}\left(R_1^\sigma, \ldots, R_n^\sigma\right)} \\ &= \frac{a_0 + \bar{a}_0 \bar{r}^n}{\sum_{i=0}^{n} a_i \bar{r}^i}.\end{aligned}$$

This measure is functionally dependent on the reliability parameter r. Clearly, our pre-theoretic notion of the coherence of an information set does not encompass the reliability of the witnesses that provide us with its content. So how can we use this measure to assess the relative coherence of two information sets?

Let us look at what we did in the special case in which information sets S and S′ have the same cardinality and $P(R_1, \ldots, R_n) = a_0 = a'_0 = P(R'_1, \ldots, R'_n)$. We salvaged the core of Bayesian Coherentism by imposing an ordering on a pair of information sets if and only if the curves representing the posterior probabilities that the contents of the information sets are true as a function of r do not criss-cross. Formally, $S \succeq S'$ if and only if $P^*(R_1, \ldots, R_n) \geq P^*(R'_1, \ldots, R'_n)$ for all values of the reliability parameter $r \in (0, 1)$. This permitted us to respect the first tenet of Bayesian Coherentism—viz. the more coherent an information set is, the greater our degree of confidence that its content is true, ceteris paribus—while remaining faithful to a weakened version of the second tenet—viz. that the quasi-ordering of *being no less coherent than* is determined by the probabilistic features of the information set.

In the general case, we would like to be able to assess and compare the coherence of information sets that may not have the same cardinality and may not share the same joint prior probability that their respective contents are true. Our strategy is to assess the coherence of an information set by measuring the proportion of the confidence boost that we actually receive, relative to the confidence boost that we would have received, had we received this very same information in the form of maximally coherent information. Also, in the general case we would like to be able to make the claim that the more coherent an information set is, the greater this proportional confidence boost, ceteris paribus, in which the ceteris paribus clause requires that the reliability parameter r be held constant. Now we run into precisely the same problem that we ran into before: Some pairs of information sets $\{S, S'\}$ are such that $c_r(S) > c_r(S')$ for some values of r, whereas $c_r(S') > c_r(S)$ for other values of r. To safeguard our current claim, we follow the same strategy. We impose an ordering on a pair of information sets if and only if the curves that represent the proportional confidence boosts as a function of r do not criss-cross. In formal terms,

(2.7) For all $S, S' \in \mathbf{S}$, $S \succeq S'$ iff $c_r(S) \geq c_r(S')$ for all values of the reliability parameter $r \in (0, 1)$.

This procedure induces a quasi-ordering on the set of information sets in general, whatever their cardinalities and whatever the prior joint probabilities that their contents are true. We will see that this distinction squares with our willingness to make intuitive judgements about the relative coherence of information sets.

The reader may wonder whether our general-case procedure entails our special-case procedure. The answer is straightforward. In the special case, we assume that the cardinalities of the information sets are equal and that the prior probabilities that the contents of the information sets are true are equal—i.e. $a_0 = a'_0$. From (2.2) and (2.6), it follows that we can write the posterior joint probability that the content of the information set is true as follows:

$$P^*(\mathrm{R}_1, \ldots, \mathrm{R}_n) = \frac{a_0}{a_0 + \bar{a}_0 \bar{r}^n} c_r(\{\mathrm{R}_1, \ldots, \mathrm{R}_n\}). \tag{2.8}$$

It is clear from (2.8) that

(2.9) For all S, $S' \in \mathbf{S}$, if S has cardinality m and S' has cardinality n with $m = n$ and $a_0 = a'_0$, then $P^*(\mathrm{R}_1, \ldots, \mathrm{R}_m) \geq P^*(\mathrm{R}'_1, \ldots, \mathrm{R}'_n)$ if and only if $c_r(\mathrm{S}) \geq c_r(\mathrm{S}')$ for all values of the reliability parameter $r \in (0, 1)$.

Our procedure in the general case, as expressed in (2.7), in conjunction with (2.9) entails our procedure in the special case, as expressed in (2.1).

Rather than assessing directly whether the curves criss-cross for the functions that measure the proportional confidence boost, we construct a *difference function*. Consider two information sets $\mathrm{S} = \{\mathrm{R}_1, \ldots, \mathrm{R}_m\}$ and $\mathrm{S}' = \{\mathrm{R}'_1, \ldots, \mathrm{R}'_n\}$. We calculate the weight vectors $< a_0, \ldots, a_m >$ and $< a'_0, \ldots, a'_n >$. The difference function is defined as follows:

$$f_r(\mathrm{S}, \mathrm{S}') = c_r(\mathrm{S}) - c_r(\mathrm{S}'). \tag{2.10}$$

$f_r(\mathrm{S}, \mathrm{S}')$ has the same sign for all values of $r \in (0, 1)$ if and only if the measure $c_r(\mathrm{S})$ is always greater than or is always smaller than the measure $c_r(\mathrm{S}')$ for all values of $r \in (0, 1)$. Hence, we can restate the general procedure in (2.7) that induces a quasi-ordering over an unrestricted set of information sets in a more parsimonious fashion:

(2.11) For two information sets $\mathrm{S}, \mathrm{S}' \in \mathbf{S}$, $\mathrm{S} \succeq \mathrm{S}'$ iff $f_r(\mathrm{S}, \mathrm{S}') \geq 0$ for all values of $r \in (0, 1)$.

If the information sets S and S′ are of equal size, then it is also possible to determine whether there exists a coherence ordering over these sets *directly* from the weight vectors $< a_0, \ldots, a_n >$ and $< a'_0, \ldots, a'_n >$. One need only evaluate the conditions under which the sign of the difference function is invariable for all values of $r \in (0, 1)$. In Appendix B.1, we have shown that

(2.12) $a'_i/a_i \geq \max(1, a'_0/a_0),\ \forall i = 1, \ldots, n-1$
is a necessary and sufficient condition for $S \succeq S'$ for $n = 2$ and is a sufficient condition for $S \succeq S'$ for $n > 2$.

This is the more parsimonious statement of the condition. However, it is easier to interpret this condition when stated as a disjunction:

(2.13) (i) $a'_0 \leq a_0$ & $a'_i \geq a_i,\ \forall i = 1, \ldots, n-1$, or,
(ii) $a'_0 \geq a_0$ & $a'_i/a_i \geq a'_0/a_0,\ \forall i = 1, \ldots, n-1$,
is a necessary and sufficient condition for $S \succeq S'$ for $n = 2$ and is a sufficient condition for $S \succeq S'$ for $n > 2$.

It is easy to see that (2.12) and (2.13) are equivalent.[1]

Let us now interpret (2.13). For $n = 2$, let $S = \{R_1, R_2\}$ and consider the diagram for the joint probability distribution in Figure 2.1. There are precisely two ways to decrease[2] the coherence in moving from information sets S to S′: First, by shrinking the overlapping area between R_1 and R_2 ($a'_0 \leq a_0$) and expanding the non-overlapping area ($a'_1 \geq a_1$); and second, by expanding the overlapping area ($a'_0 \geq a_0$) while expanding the non-overlapping area to a greater degree ($a'_1/a_1 \geq a'_0/a_0$). The example of the corpse in Tokyo in the next section is meant to show that these conditions are intuitively plausible.

For $n > 2$, consider the diagram for the joint probability distribution in Figure 2.2 and let $S = \{R_1, R_2, R_3\}$. There are two ways to decrease the coherence in moving from S to S′: First, by shrinking the area in which there is complete overlap between $R_1, \ldots, R_n$ ($a'_0 \leq a_0$)

[1] Assume (2.12). Either $max(1, a'_0/a_0) = 1$ or $max(1, a'_0/a_0) = a'_0/a_0$. In the former case, it follows from the inequality in (2.12) that $a'_0 \leq a_0$ and $a'_i \geq a_i,\ \forall i = 1, \ldots, n-1$. In the latter case, it follows from the inequality in (2.12) that $a'_0 \geq a_0$ and $a'_i/a_i \geq a'_0/a_0,\ \forall i = 1, \ldots, n-1$. Hence, (2.13) follows. Assume (2.13). Suppose (i) holds. From the first conjoint in (i), $max(1, a'_0/a_0) = 1$ and hence from the second conjoint in (i), $a'_i/a_i \geq max(1, a'_0/a_0), \forall i = 1, \ldots, n-1$. Suppose (ii) holds. From the first conjoint in (ii), $max(1, a'_0/a_0) = a'_0/a_0$ and hence from the second conjoint in (ii), $a'_i/a_i \geq max(1, a'_0/a_0), \forall i = 1, \ldots, n-1$. Hence, (2.12) follows.

[2] We introduce the convention that 'decreasing' stands for *decreasing or not changing*, 'shrinking' for *shrinking or not changing*, and 'expanding' for *expanding or not changing*. This convention permits us to state the conditions in (2.13) more clearly and is analogous to the microeconomic convention to let 'preferring' stand for *weak* preference, i.e. for *preferring to or being indifferent between* in ordinary language.

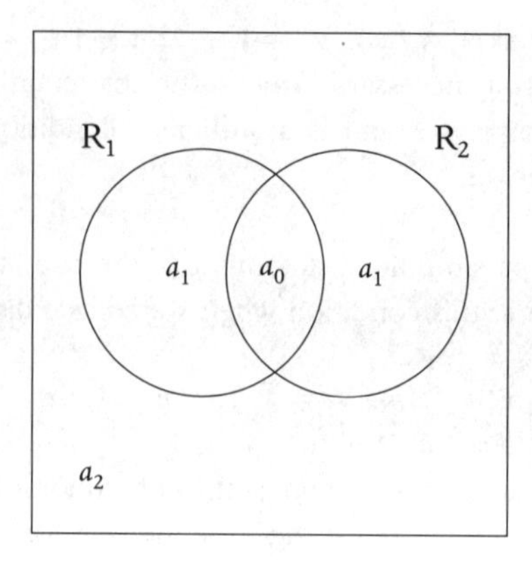

FIG. 2.1 A diagram for the probability distribution for information pairs

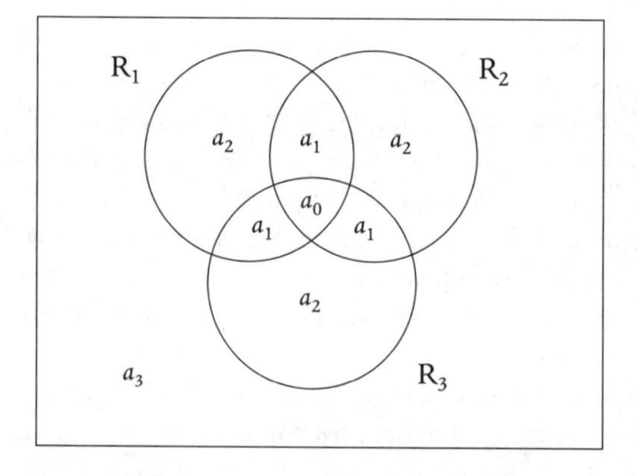

FIG. 2.2 A diagram for the probability distribution for information triples

and expanding all the areas in which there is no complete overlap ($a_i' \geq a_i$, $\forall i = 1, \ldots, n-1$); and second, by expanding the area in which there is complete overlap ($a_0' \geq a_0$) and expanding all the non-overlapping areas to a greater degree ($a_i'/a_i \geq a_0'/a_0$, $\forall i = 1, \ldots, n-1$). This is a sufficient but not a necessary condition for $n > 2$. Hence, if equal-sized information sets do not satisfy condition (2.13), we still need to apply our general method in (2.11), i.e. we need to examine the sign of $f_r(S, S')$ for all values of $r \in (0, 1)$. The example of BonJour's challenge in the next section shows that it may be possible to order two information sets using the general method in (2.11) without satisfying the sufficient condition in (2.13).

If we wish to determine the relative coherence of two information sets S and S′ of unequal size, we have no shortcut. In that case, we need to apply our general method in (2.11), i.e. we need to examine the sign of $f_r(S, S')$ for all values of $r \in (0, 1)$. The example of Tweety in the next section will provide an illustration of the procedure used to judge the relative coherence of information sets of unequal size.

2.3. A CORPSE IN TOKYO, BONJOUR'S RAVENS AND TWEETY

Does our analysis yield the correct results for some intuitively clear cases? We consider a comparison (i) of two information *pairs*, (ii) of two information *triples*, and (iii) of two information sets of unequal size.

(i) Information Pairs. Suppose that we are trying to locate a corpse from a murder somewhere in Tokyo. We draw a grid of 100 squares over the map of the city and consider it equally probable that the corpse lies somewhere within each square. We interview two partially and equally reliable witnesses. Suppose witness 1 reports that the corpse is somewhere in squares 50 to 60 and witness 2 reports that the corpse is somewhere in squares 51 to 61. Call this situation α and include this information in the information set S^{α}. For this information set, $a_0^{\alpha} = .10$ and $a_1^{\alpha} = .02$.

Let us now consider a different situation in which the reports from the two sources overlap far less. In this alternate situation—call it β—witness 1 reports squares 20 to 55 and witness 2 reports squares 55 to 90. This information is contained in S^{β}. The overlapping area shrinks to $a_0^{\beta} = .01$ and the non-overlapping area expands to $a_1^{\beta} = .70$. On condition (2.13)(i), S^{β} is less coherent than S^{α}, since $a_0^{\beta} = .01 \leq a_0^{\alpha} = .10$ and $a_1^{\beta} = .70 \geq a_1^{\alpha} = .02$.

In a third situation γ, witness 1 reports squares 20 to 61 and witness 2 reports squares 50 to 91. S^{γ} contains this information. The overlapping area expands to $a_0^{\gamma} = .12$ and the non-overlapping area expands to $a_1^{\gamma} = .60$. On condition (2.13)(ii), S^{γ} is less coherent than S^{α}, since $a_0^{\gamma} = .12 \geq a_0^{\alpha} = .10$ and $a_1^{\gamma}/a_1^{\alpha} = 30 \geq 1.2 = a_0^{\gamma}/a_0^{\alpha}$.

Now let us consider a pair of situations in which no ordering of the information sets is possible. We are considering information pairs, i.e. $n = 2$, and so condition (2.12) and (2.13) provide equivalent necessary and sufficient conditions to order two information pairs, *if there exists an ordering*. In situation δ, witness 1 reports squares 41 to 60 and witness 2 reports squares 51 to 70. So $a_0^\delta = .10$ and $a_1^\delta = .20$. In situation ε, witness 1 reports squares 39 to 61 and witness 2 reports squares 50 to 72. So $a_0^\varepsilon = .12$ and $a_1^\varepsilon = .22$. Is the information set in situation δ more or less coherent than in situation ε? It is more convenient here to invoke condition (2.12). Notice that $a_1^\varepsilon / a_1^\delta = 1.10$ is not greater than or equal to $1.20 = max\left(1, a_0^\varepsilon / a_0^\delta\right)$, nor is $a_1^\delta / a_1^\varepsilon \approx .91$ greater than or equal to $1 = max\,(1, a_0^\delta / a_0^\varepsilon)$. Hence neither $S^\delta \succeq S^\varepsilon$ nor $S^\varepsilon \succeq S^\delta$ hold true.

These quasi-orderings over the information sets in situations α and β, in situations α and γ, and in situations δ and ε seems to square quite well with our intuitive judgements. Without having done any empirical research, we conjecture that most experimental subjects would indeed rank the information set in situation α to be more coherent than the information sets in either situations β or γ. Furthermore, we also conjecture that if one were to impose sufficient pressure on the subjects to judge which of the information sets in situations δ and ε is more coherent, we would be left with a split vote.

We have reached these results by applying the special conditions in (2.12) and (2.13) for comparing information sets. The same results can be obtained by using the general method in (2.11). Write down the difference functions as follows for each comparison (i.e. let $i = \alpha$ and $j = \beta$, let $i = \alpha$ and $j = \gamma$, and let $i = \delta$ and $j = \varepsilon$ in turn):

$$(2.14)\quad f_r(S^i, S^j) = c_r(S^i) - c_r(S^j) = \frac{a_0^i + \bar{a}_0^i \bar{r}^2}{a_0^i + a_1^i \bar{r} + a_2^i \bar{r}^2} - \frac{a_0^j + \bar{a}_0^j \bar{r}^2}{a_0^j + a_1^j \bar{r} + a_2^j \bar{r}^2}.$$

As we can see in Figure 2.3, the functions $f_r(S^\alpha, S^\beta)$ and $f_r(S^\alpha, S^\gamma)$ are positive for all values of $r \in (0, 1)$—so S^α is more coherent than S^β and S^γ. But $f_r(S^\delta, S^\varepsilon)$ is positive for some values and negative for other values of $r \in (0, 1)$—so there is no coherence ordering over S^δ and S^ε.

(ii) Information Triples. We return to BonJour's challenge. There is a more coherent set, $S = \{R_1 =$ [All ravens are black], $R_2 =$ [This bird is a raven], $R_3 =$ [This bird is black]$\}$, and a less coherent set, $S' = \{R_1' =$ [This chair is brown], $R_2' =$ [Electrons are negatively charged],

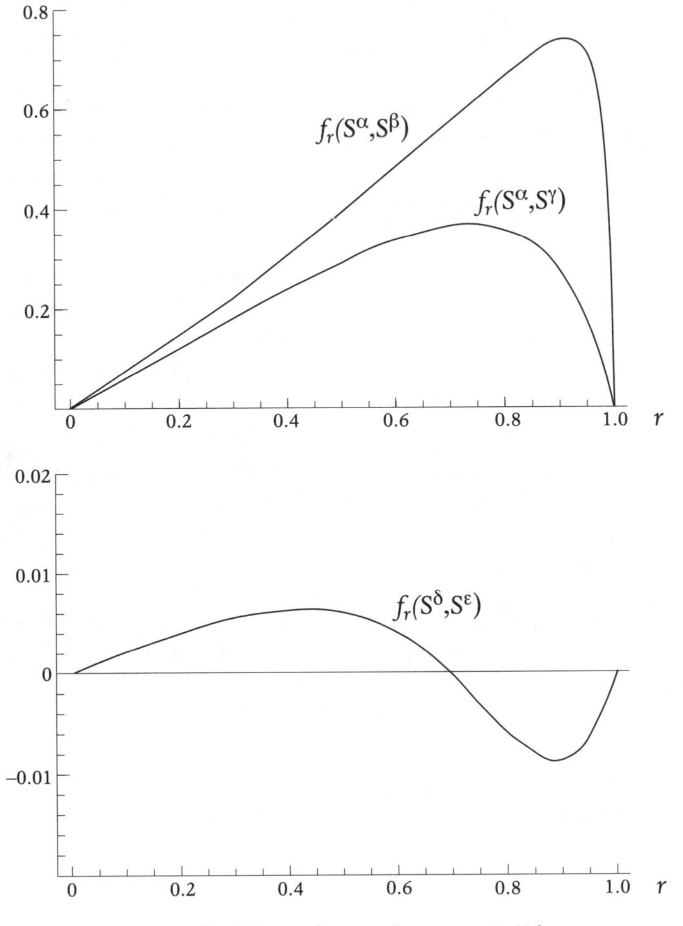

FIG. 2.3 The difference functions for a corpse in Tokyo

$R_3' =$ [Today is Thursday]}. The challenge is to give an account of the fact that S is more coherent than S′. Let us apply our analysis to this challenge.

What is essential in S is that $R_1\&R_2 \vdash R_3$, so that $\boldsymbol{P}(R_3|R_1, R_2) = 1$. But to construct a joint probability distribution, we need to make some additional assumptions. Let us make assumptions that could plausibly describe the degrees of confidence of an amateur ornithologist who is sampling a population of birds:

(i) There are four species of birds in the population of interest, ravens being one of them. There is an equal chance of picking a bird from each species: $\boldsymbol{P}(\mathrm{R}_2) = 1/4$.

(ii) The random variables R_1 and R_2, whose values are the propositions R_1 and $\neg\mathrm{R}_1$, and R_2 and $\neg\mathrm{R}_2$, respectively, are probabilistically independent: Learning no more than that a raven was (or was not) picked teaches us nothing at all about whether all ravens are black.

(iii) We have prior knowledge that birds of the same species *often* have the same colour and black may be an appropriate colour for a raven. Let us set $\boldsymbol{P}(\mathrm{R}_1) = 1/4$.

(iv) There is a one in four chance that a black bird has been picked amongst the non-ravens, whether all ravens are black or not, i.e. $\boldsymbol{P}(\mathrm{R}_3|\neg\mathrm{R}_1, \neg\mathrm{R}_2) = \boldsymbol{P}(\mathrm{R}_3|\mathrm{R}_1, \neg\mathrm{R}_2) = 1/4$. Since we know that birds of a single species often share the same colour, there is only a chance of $1/10$ that the bird that was picked happens to be black, given that it is a raven and that it is not the case that all ravens are black, i.e. $\boldsymbol{P}(\mathrm{R}_3|\neg\mathrm{R}_1, \mathrm{R}_2) = 1/10$.

These assumptions permit us to construct the joint probability distribution for R_1, R_2, R_3 and to specify the weight vector $<a_0, \ldots, a_3>$ (see Figure 2.4).[3]

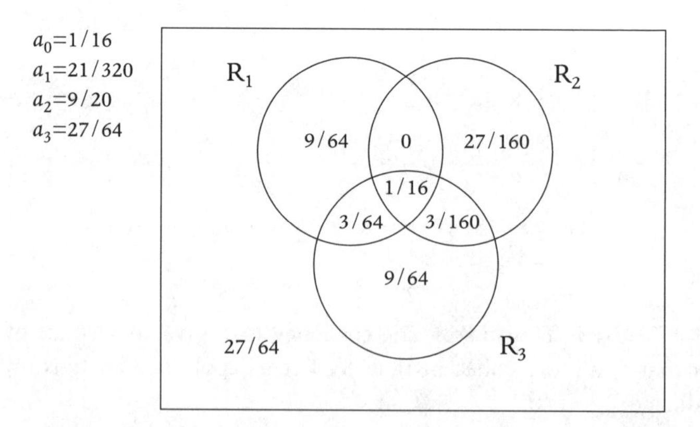

FIG. 2.4 A diagram for the probability distribution for the set of dependent propositions in BonJour's ravens

[3] Since R_1 and R_2 are probabilistically independent, $\boldsymbol{P}(R_1, R_2, R_3) = \boldsymbol{P}(R_1)\boldsymbol{P}(R_2)\boldsymbol{P}(R_3|R_1, R_2)$ for all values of R_1, R_2, and R_3. The numerical values in Figure 2.4 can be directly calculated.

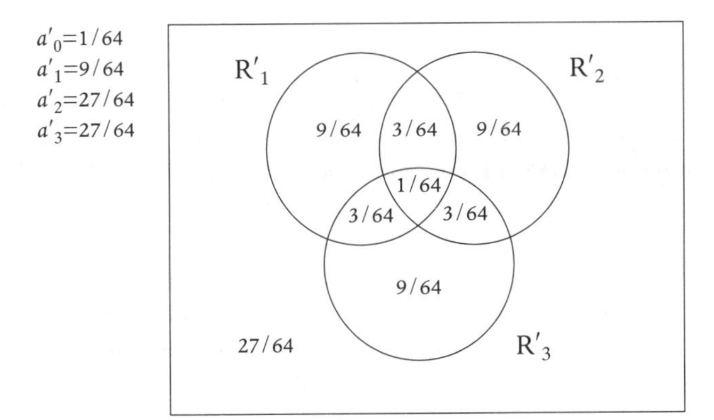

FIG. 2.5 A diagram for the probability distribution for the set of independent propositions in BonJour's ravens

What is essential in information set S′ is that the propositional variables are probabilistically independent—e.g. learning something about electrons presumably does not teach us anything about what day it is today or about the colour of a chair. Let us suppose that the marginal probabilities of each proposition are $\mathbf{P}(\mathrm{R}'_1) = \mathbf{P}(\mathrm{R}'_2) = \mathbf{P}(\mathrm{R}'_3) = 1/4$. We construct the joint probability distribution for R'_1, R'_2, and R'_3 and specify the weight vector $<a'_0, \ldots, a'_3>$ in Figure 2.5.[4]

The information triples do not pass the *sufficient* condition for the determination of the direction of the coherence ordering in (2.12).[5] So we need to appeal to our general method and construct the difference function:

$$\text{(2.15)} \quad f_{ravens} = f_r(\mathrm{S}, \mathrm{S}') = \frac{a_0 + \bar{a}_0\bar{r}^3}{a_0 + a_1\bar{r} + a_2\bar{r}^2 + a_3\bar{r}^3} - \frac{a'_0 + \bar{a}'_0\bar{r}^3}{a'_0 + a'_1\bar{r} + a'_2\bar{r}^2 + a'_3\bar{r}^3}.$$

We have plotted f_{ravens} in Figure 2.7. This function is positive for all values of $r \in (0, 1)$. Hence we may conclude that S is more coherent than S′, which is precisely the intuition of which BonJour wanted an account.[6]

[4] Since R'_1, R'_2, and R'_3 are probabilistically independent, $\mathbf{P}(R'_1, R'_2, R'_3) = \mathbf{P}(R'_1)\mathbf{P}(R'_2)\mathbf{P}(R'_3)$ for all values of R'_1, R'_2, and R'_3. The numerical values in Figure 2.5 can be directly calculated.

[5] Clearly the condition fails for $\mathrm{S}' \succeq \mathrm{S}$, but it also fails for $\mathrm{S} \succeq \mathrm{S}'$, since $a'_2/a_2 \approx .94 < 1 = max\,(1, .25) = max\,(1, a'_0/a_0)$.

[6] It is not always the case that an information triple in which one of the propositions is entailed by the two other propositions is more coherent than an information triple in

(iii) Information Sets of Unequal Size. Finally, we consider a comparison between an information pair and an information triple. The following example is inspired by the paradigmatic example of non-monotonic reasoning about Tweety the penguin. We are not interested in non-monotonic reasoning here, but merely in the question of the coherence of information sets. Suppose that we come to learn from independent sources that someone's pet Tweety is a bird (B) and that Tweety cannot fly, i.e. that Tweety is a ground-dweller (G). Considering what we know about pets, {B, G} is highly incoherent information. Aside from the occasional penguin, there are no ground-dwelling birds that qualify as pets, and aside from the occasional bat, there are no flying non-birds that qualify as pets. Later, we receive the new item of information that Tweety is a penguin (P). Our extended information set $S' = \{B, G, P\}$ seems to be much more coherent than $S = \{B, G\}$. So let us see whether our analysis bears out this intuition. We construct a joint probability distribution for *B*, *G*, and *P* together with the marginalized probability distributions for *B* and *G* in Figure 2.6.

Since the information sets are of unequal size, we need to appeal to our general method in (2.11) and construct the difference function:

$$(2.16) \quad f_{tweety} = f_r(S', S) = \frac{a'_0 + \bar{a}'_0\bar{r}^3}{a'_0 + a'_1\bar{r} + a'_2\bar{r}^2 + a'_3\bar{r}^3} - \frac{a_0 + \bar{a}_0\bar{r}^2}{a_0 + a_1\bar{r} + a_2\bar{r}^2}.$$

We have plotted f_{tweety} in Figure 2.7. This function is positive for all values of $r \in (0,1)$. We may conclude that S′ is more coherent than S, which is precisely the intuition that we wanted to account for.

which the propositions are probabilistically independent. For instance, suppose that R_2 and R_3 are extremely incoherent propositions, i.e. the truth of R_2 makes R_3 extremely implausible and vice versa, and that R_1 is an extremely implausible proposition which in conjunction with R_2 entails R_3. Then it can be shown that this set of propositions is not a more coherent set than a set of probabilistically independent propositions. This is not unwelcome, since entailments by themselves should not warrant coherence. Certainly, $\{R_1, R_2, R_3\}$ should not be a coherent set when R_2 and R_3 are inconsistent and R_1 contradicts our background knowledge, although $R_1 \& R_2 \vdash R_3$. A judgement to the effect that S is more coherent than S′ depends both on logical relationships and background knowledge.

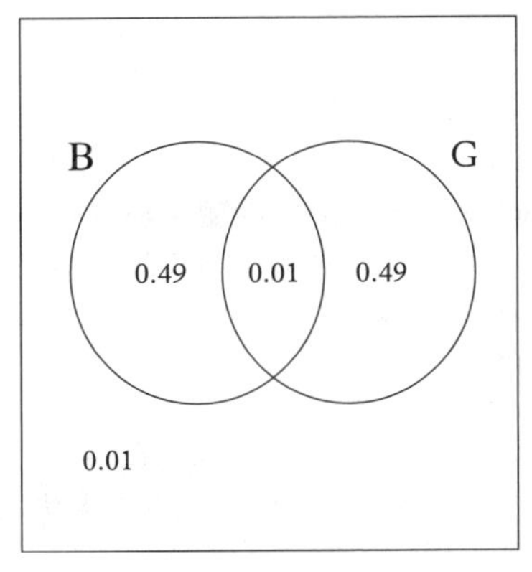

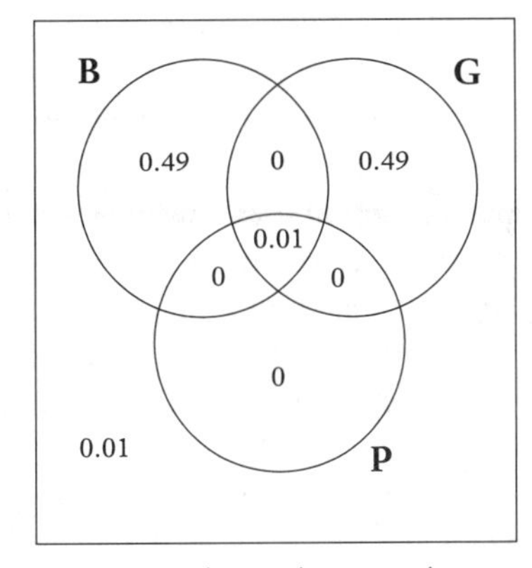

a_0=0.01; a_1=0.98; a_2=0.01　　　　a_0'=0.01; a_1'=0; a_2'=0.98; a_3'=0.01

Fig. 2.6 A diagram for the probability distribution for Tweety before and after extension with [Tweety is a penguin]

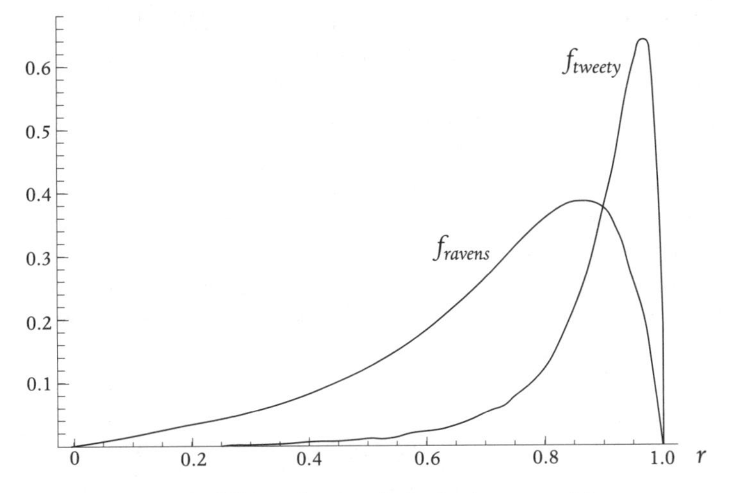

Fig. 2.7 The difference functions for BonJour's ravens and Tweety

2.4. EQUAL RELIABILITY

We have built into our model the assumption that the sources are equally reliable, i.e. that all sources have the same true positive rate p and the same false positive rate q. This seems like an unreasonably strong assumption, since, when we are gathering information in the actual world, we typically trust some sources less and some sources more. But our assessment of the relative coherence of information sets has nothing to do with how much we *actually* trust our information sources. As a matter of fact, we may assess the coherence of an information set without having any clue whatsoever who the sources are of the items in this information set or what their degrees of reliability are. An assessment of coherence requires a certain metric that features *hypothetical* sources with certain idealized characteristics. These hypothetical sources are not epistemically perfect, as is usually the case in idealizations. Rather, they are characterized by idealized *imperfections*—their partial reliability. Furthermore, our idealized sources possess the same degree of *internal* reliability and the same degree of *external* reliability. By internal reliability we mean that the sources for each item within an information set are equally reliable, and by external reliability we mean that the sources for each information set are equally reliable.

To see why internal reliability is required in our model, consider the following two information sets. Set S contains two equivalent propositions R_1 and R_2 and a third proposition R_3 that is highly negatively relevant with respect to R_1 and R_2. Set S' contains three propositions R'_1, R'_2, and R'_3 and every two propositions in S' are just short of being equivalent. One can specify the contents of such information sets such as to make S' intuitively more coherent than S. Our formal analysis will agree with this intuition. Now suppose that it turns out that the actual—i.e. the non-idealized—information sources for R_1, R'_1, R_2, and R'_2 are quite reliable and for R_3 and R'_3 are close to fully unreliable. We assign certain values to the reliability parameters to reflect this situation and calculate the proportional confidence boosts that actually result for both information sets. Plausible values can be picked for the relevant parameters so that the proportional confidence boost for S actually *exceeds* the proportional confidence boost for S'. This comes about because the actual information sources virtually bring nothing to the propositions R_3 and R'_3 and because R_1 and R_2 are indeed equivalent (and hence maximally coherent), whereas R'_1 and R'_2 are short of being equivalent (and hence less than maximally coherent). But what we want is an assessment of the relative coherence of $\{R_1, R_2, R_3\}$ and $\{R'_1, R'_2, R'_3\}$ and not of the relative coherence of $\{R_1, R_2\}$ and $\{R'_1, R'_2\}$. The appeal to ideal agents with the same degree of internal reliability in our metric is warranted by the fact that we want to compare the degree of coherence of complete information sets and not of some proper subsets of them.

Second, to see why *external* reliability is required in our model, consider some information set S which is not maximally coherent, but clearly more coherent than an information set S′. Any of our examples in Section 2.3 will do for this purpose. It is always possible to pick two values r and r' so that $c_{r'}(S') > c_r(S)$. To obtain such a result, we need only pick a value of r' in the neighbourhood of 0 or 1 and pick a less extreme value for r, since it is clear from (2.6) that for r' approaching 0 or 1, $c_{r'}(S')$ approaches 1. This is why coherence needs to be assessed relative to idealized sources that are taken to have the same degree of external reliability.

2.5. INDETERMINACY

Our analysis has some curious repercussions for the indeterminacy of comparative judgements of coherence. Consider the much debated problem among Bayesians of how to set the prior probabilities. We have chosen examples in which shared background knowledge (or ignorance) imposes constraints on what prior joint probability distributions are reasonable.[7] In the case of the corpse in Tokyo, one could well imagine coming to the table with no prior knowledge whatsoever about where an object is located in a grid with equal-sized squares. Then it seems reasonable to assume a uniform distribution over the squares in the grid. In the case of BonJour's ravens we modelled a certain lack of ornithological knowledge and let the joint probability

[7] Note that this is no more than a framework of presentation. Our approach is actually neutral when it comes to interpretations of probability. Following Gillies (2000), we favour a pluralistic view of interpretations of probability. The notion used in a certain context depends on the application in question. But, if one believes, as a more zealous personalist, that only the Kolmogorov axioms and Bayesian updating impose constraints on what constitute reasonable degrees of confidence, then there will be less room for rational argument and intersubjective agreement about the relative coherence of information sets. Or, if one believes, as an objectivist, that joint probability distributions can only be meaningful when there is the requisite objective ground, then there will be less occasion for comparative coherence judgements. None of this affects our project. The methodology for the assessment of the coherence of information sets remains the same, no matter what interpretation of probability one embraces.

distribution respect the logical entailment relation between the propositions in question. In the case of Tweety one could make use of frequency information about some population of pets that constitutes the appropriate reference class.

But often we find ourselves in situations without such reasonable constraints. What are we to do then? For instance, what is the probability that the butler was the murderer (B), given that the murder was committed with a kitchen knife (K), that the butler was having an affair with the victim's wife (A), and that the murderer was wearing a butler jacket (J)? Certainly the prior joint probability distributions over the propositional variables *B*, *K*, *A*, and *J* may reasonably vary widely for different Bayesian agents and there is little that we can point to in order to adjudicate in this matter. But to say that there is room for legitimate disagreement among Bayesian agents is not to say that anything goes. Certainly we will want the joint probability distributions to respect, among others things, the feature that $\boldsymbol{P}(\mathrm{B}|\mathrm{K},\mathrm{A},\mathrm{J}) > \boldsymbol{P}(\mathrm{B})$. Sometimes there are enough rational constraints on degrees of confidence to warrant agreement in comparative coherence judgements over information sets. And sometimes there are not. It is perfectly possible for two rational agents to have degrees of confidence that are so different that they are unable to reach agreement about comparative coherence judgements. This is one kind of indeterminacy. Rational argument cannot always bring sufficient precision to degrees of confidence to yield agreement on judgements of coherence.

But what our analysis shows is that this is not the only kind of indeterminacy. Two rational agents may have the same subjective joint probability distribution over the relevant propositional variables and still be unable to make a comparative judgement about two information sets. This is so for situations δ and ε in the case of the corpse in Tokyo. Although there is no question about what constitutes the proper joint probability distributions that are associated with the information sets in question, no comparative coherence judgement about S^{δ} and S^{ε} is possible. This is so because the proportional confidence boost for S^{δ} exceeds the proportional confidence boost for S^{ε} for some intervals of the reliability parameter, and vice versa for other intervals. If coherence is to be measured by the proportional confidence boost and if it is to be independent of the reliability of the witnesses, then there will not exist a coherence ordering for some pairs of information sets.

In short, indeterminacy about coherence may come about because rationality does not sufficiently constrain the relevant degrees of confidence. In this case, it is our epistemic predicament with respect to the content of the information set that is to blame. However, even when the probabilistic features of a pair of information sets are fully transparent, it may still fail to be the case that one information set is more coherent than (or equally coherent as) the other. Prima facie judgements can be made on both sides, but no judgement *tout court* is warranted. In this case, indeterminacy is not due to our epistemic predicament, but rather to the probabilistic features of the information sets.

2.6. ALTERNATIVE PROPOSALS

We return to the alternative proposals to construct a coherence ranking that were introduced in Chapter 1 and will show that these proposals yield counter-intuitive results. First, Lewis does not propose a measure that induces an ordering over information sets. Rather, he claims that coherent (or, in his words, congruent) information sets have the following property

$$(2.17) \quad \boldsymbol{P}(\mathrm{R_i}|\mathrm{R_1}, \ldots, \mathrm{R_{i-1}}, \mathrm{R_{i+1}}, \ldots, \mathrm{R_n}) > \boldsymbol{P}(\mathrm{R_i}) \text{ for all } i = 1, \ldots, n.$$

But let us suppose that an information set contains n pairs of equivalent propositions, but that there is a relation of strong negative relevance (but not of inconsistency) between the propositions in each pair and all other propositions. In other words, $\boldsymbol{P}(\mathrm{R_i}, \mathrm{R_j}) > \boldsymbol{P}(\mathrm{R_i}, \mathrm{R_j}| \mathrm{R_1}, \ldots, \mathrm{R_{i-1}}, \mathrm{R_{i+1}}, \ldots, \mathrm{R_{j-1}}, \mathrm{R_{j+1}}, \ldots, \mathrm{R_{2n}}) \approx 0$ but not equal to 0, for each equivalent pair of propositions $\{\mathrm{R_i}, \mathrm{R_j}\}$. Then one would be hard-pressed to say that this information set is coherent. And yet, according to Lewis, this information set is coherent, because, assuming non-extreme marginal probabilities, $1 = \boldsymbol{P}(\mathrm{R_i}|\mathrm{R_1}, \ldots, \mathrm{R_{i-1}}, \mathrm{R_{i+1}}, \ldots, \mathrm{R_n}) > \boldsymbol{P}(\mathrm{R_i})$ for all $i = 1, \ldots, 2n$.[8]

Second, Shogenji proposes that

$$(2.18) \qquad \mathrm{S} \succeq \mathrm{S'} \text{ iff } m_s(\mathrm{S}) = \frac{\boldsymbol{P}(\mathrm{R_1}, \ldots, \mathrm{R_m})}{\prod_{i=1}^{n} \boldsymbol{P}(\mathrm{R_i})} \geq \frac{\boldsymbol{P}(\mathrm{R'_1}, \ldots, \mathrm{R'_n})}{\prod_{i=1}^{n} \boldsymbol{P}(\mathrm{R'_i})} = m_s(\mathrm{S'}).$$

[8] For an example, see Bovens and Olsson (2000: 688–9).

The following example shows that the Shogenji measure is counter-intuitive. Suppose that there are 1,000 equiprobable suspects for a crime with equal proportions of Africans, North Americans, South Americans, Europeans, and Asians. Now consider the information sets $S = \{R_1 =$ [The culprit is either an African, a North American, a South American, or a European], $R_2 =$ [The culprit is not Asian]$\}$ and $S' = \{R'_1 =$ [The culprit is an African], $R'_2 =$ [The culprit is either Youssou (a particular African), Sulla (a particular South American), or Pierre (a particular European)]$\}$. Since S contains propositions that pick out coextensive sets of suspects, whereas there is relatively little overlap between the propositions in S′, it seems reasonable to say that S is a more coherent set than S′. However, on the Shogenji measure, $m_s(S) = \frac{.8}{.8 \times .8} = 1.25 < 1.67 = \frac{.001}{.2 \times .003} = m_s(S')$. Our procedure, on the other hand, clearly matches the intuitive result in this case. The proportional confidence boost measure c_r is maximal for the maximally coherent information set S containing equivalent propositions. Hence, the difference function $f_r(S, S') = c_r(S) - c_r(S') > 0$ for all values of $r \in (0, 1)$ and so, by (2.11), S is more coherent than S′.

Third, Olsson tentatively proposes that

$$(2.19) \qquad S \succeq S' \text{ iff}$$

$$m_o(S) = \frac{P(R_1, \ldots, R_m)}{P(R_1 \vee \ldots \vee R_m)} \geq \frac{P(R'_1, \ldots, R'_n)}{P(R'_1 \vee \ldots \vee R'_n)} = m_o(S').$$

The Tweety example shows that this measure is counter-intuitive. It seems reasonable to say that the information *pair* S = {[My pet Tweety is a bird], [My pet Tweety cannot fly]} is less coherent than the information *triple* S′ = {[My pet Tweety is a bird], [My pet Tweety cannot fly], [My pet Tweety is a penguin]}. But from Figure 2.6 we can read off that $m_o(S) = .01/.99 = m_o(S')$.

Fourth, we focus on Fitelson's measure as applied to information *pairs*. The Kemeny–Oppenheim measure is a measure of factual support when the marginal probabilities of R_1 and R_2 are not extreme:

$$(2.20) \qquad F(R_1, R_2) = \frac{P(R_1|R_2) - P(R_1|\neg R_2)}{P(R_1|R_2) + P(R_1|\neg R_2)}$$

$$\text{for } P(R_1) < 1 \text{ and } P(R_2) > 0.$$

Fitelson proposes that

(2.21) $S \succeq S'$ iff

$$m_f(S) = \frac{F(R_1, R_2) + F(R_2, R_1)}{2} \geq \frac{F(R'_1, R'_2) + F(R'_2, R'_1)}{2} = m_f(S').$$

The following example shows that this measure yields counter-intuitive results. Let there be 100 suspects for a crime who have an equal chance of being the culprit. In situation one, let there be 6 Trobriand suspects and 6 chess-playing suspects; there is 1 Trobriand chess player. In situation two, let there be 85 Ik suspects and 85 rugby-playing suspects; there are 80 Ik rugby players. Which information is more coherent—$S = \{R_1 =$ [The culprit is a Trobriand], $R_2 =$ [The culprit is a chess player]$\}$ or $S' = \{R'_1 =$ [The culprit is an Ik], $R'_2 =$ [The culprit is a rugby player]$\}$? The information in S' seems to fit together much better than in S, since there is so little overlap between being a Trobriander and being a chess player and there is considerable overlap between being an Ik and a rugby player. But note that on Fitelson's measure $m_f(S) \approx .52 > .48 \approx m_f(S')$. The Fitelson measure behaves curiously for cases in which we increase the overlapping area, while keeping the non-overlapping area fixed. Intuitively, one would think that when keeping the non-overlapping area fixed, then, the more overlap, the greater the coherence. And this is indeed what our condition (2.12) indicates. But on the Fitelson measure, this is not the case. In Figure 2.8, we set the non-overlapping area at $P(R_1, \neg R_2) = P(\neg R_1, R_2) = .05$. We increase the overlapping area a_0 from .01 to .80 and plot the Fitelson measure as a function of a_0 in Figure 2.9. The measure first increases from $a_0 = .01$ and then reaches its maximum for $a_0 \approx .17$ and subsequently decreases again. We fail to see any intuitive justification for this behaviour of the measure.

Where do these proposals go wrong? Lewis forgets that strong positive relevance between each proposition in a singleton set and the propositions in the complementary set is compatible with strong negative relevance between certain propositions in the information set. On Shogenji's measure, information sets containing less probable propositions tend to do better on the coherence score, so much so that information sets with non-equivalent but less probable propositions may

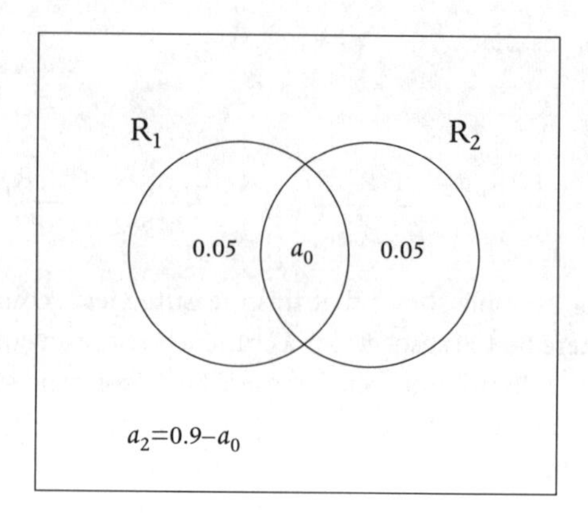

FIG. 2.8 A diagram for the probability distributions of the information sets in our counter-example to the Fitelson measure

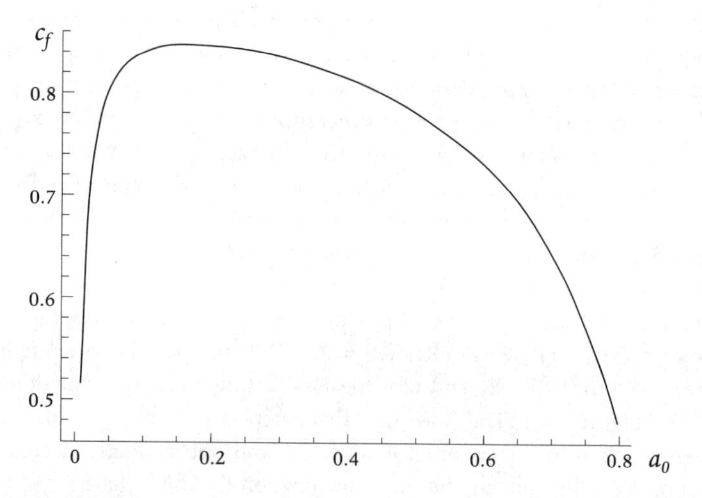

FIG. 2.9 The Fitelson measure m_f as a function of $a_0 \in [.01, .8]$ for the information sets in our counter-example to the Fitelson measure

score higher than information sets containing all and only equivalent propositions. We concur with Fitelson (2003) that an information set with all and only equivalent propositions is maximally coherent. It is not possible for non-equivalent propositions to fit together better than equivalent propositions. Certainly, information sets with less probable propositions may be more informative—it is more informative when a suspect points to Sulla than when she points to the whole group of South Americans. Furthermore, informativeness is a good-making characteristic of witness reports, as is coherence. But this is no reason to think that informativeness should be an aspect of coherence. Olsson pays exclusive attention to the relative overlap between the propositions in the information set. But note that by increasing the number of propositions one can increase relations of positive relevance while keeping the relative overlap fixed. Fitelson's measure assesses the degree of positive relevance between the propositions in the information set. But sometimes the relative overlap between the propositions gets the upper hand in our intuitive judgement of coherence.

We believe that judgements of coherence rest on the subtle interplay between the degree of positive relevance relations and relative overlap relations between propositions. To determine the nature of this subtle interplay, it is of no use to consult our intuitions. Rather, one needs to determine the relative coherence through the role that coherence is meant to play—the role of boosting our confidence in the propositions in question. More coherent information sets are information sets that display higher proportional coherence boosts regardless of the degree of reliability of the sources.

2.7. THEORY CHOICE IN SCIENCE

Where does our analysis leave the claim in philosophy of science that coherence plays a role in theory choice? We repeat the equality in (2.8):

$$P^*(\mathrm{R}_1, \ldots, \mathrm{R}_n) = \frac{a_0}{a_0 + \bar{a}_0 \bar{r}^n} \times c_r(\{\mathrm{R}_1, \ldots, \mathrm{R}_n\}). \tag{2.22}$$

What this means is that our degree of confidence in an information set S can be expressed in terms of the measure $c_r(\mathrm{S})$ which induces a quasi-ordering weighted by a factor. Note that this factor approximates 1 for larger information sets (large n) as well as for highly reliable sources ($r \approx 1$). Let us assume that we are comparing two information sets that can be ordered. Then the relative degree of confidence for these two information sets is fully determined by their relative coherence, if either the sources are sufficiently reliable or the information sets are sufficiently large.

One can represent a scientific theory T by a set of propositions $\{T_1, \ldots, T_m\}$. Let the T_is be assumptions, scientific laws, specifications of parameters, and so on. It is not plausible to claim that each proposition is independently tested, i.e. that each T_i shields off the evidence E_i for this proposition from all other propositions in the theory and all other evidence. The constitutive propositions of a theory are tested in unison. They are arranged into models that combine various propositions in the theory. Different models typically share some of their contents, i.e. some propositions in T may play a role in multiple models. It is more plausible to claim that each model M_i is being supported by some set of evidence E_i and that each M_i shields off the evidence E_i in support of the model from the other models in the theory and from other evidence. This is what it means for the models to be supported by independent evidence. There are complex probabilistic relations between the various models in the theory.

Formally, let each M_i for $i = 1, \ldots, n$ combine the relevant propositions of a theory T that are necessary to account for the independent evidence E_i. A theory T can be represented as the union of these M_is.[9] Let M_i be the variable which ranges over the value M_i stating that all propositions in the model are true and the value $\neg M_i$ stating that at least one proposition in the model is false. In Bayesian confirmation theory, E_i is evidence for M_i if and only if the likelihood ratio

$$x_i = \frac{P(E_i|\neg M_i)}{P(E_i|M_i)} \tag{2.23}$$

is contained in (0,1). Hence, E_i stands to M_i in the same way as $REPR_i$ stands to R_i in our framework. Let us suppose that all the likelihood ratios x_i equal x. $\bar{x} := 1 - x$ now plays the same role as r in our earlier model. We can construct a probability measure P for the constituent models of a theory T and identify the weight vector $<a_0, \ldots, a_n>$. If we translate the constraints of our earlier model, the following result holds up:

[9] This account of what a scientific theory is contains elements of both the syntactic view and the semantic view. Scientific theories are characterized by the set of their models, as on the semantic view, and these models (as well as the evidence for the models) are expressed as sets of propositions, as on the syntactic view.

$$P^*(\mathrm{M}_1, \ldots, \mathrm{M}_\mathrm{n}) = \frac{a_0}{a_0 + \bar{a}_0 x^n} \times c_{\bar{x}}(\{\mathrm{M}_1, \ldots, \mathrm{M}_\mathrm{n}\}). \tag{2.24}$$

Suppose that we are faced with two contending theories. The models within each theory are supported by independent items of evidence. It follows from (2.24) that, if (i) the evidence for each model is equally strong, as expressed by a single parameter x, *and*, (ii) either the evidence for each model is relatively strong ($x \approx 0$), *or*, each theory can be represented by a sufficiently large set of models (large n), then a higher degree of confidence is warranted for the theory that is represented by the more coherent set of models. Of course, we should not forget the *caveat* that indeterminacy springs from two sources. First, there may be substantial disagreement about the prior joint probability distribution over the variables $M_1, \ldots, M_n$, and second, even in the absence of such disagreement, no comparative coherence judgement may be possible between both theories, represented by their respective constitutive models. But even in the face of our assumptions and the *caveats* concerning indeterminacy, this is certainly not a trivial result about the role of coherence in theory choice within the framework of Bayesian confirmation theory.

3

Reliability

3.1. RELIABILITY DEFINED ENDOGENOUSLY

So far we have thought about partially reliable sources on the model of medical tests. There is the true positive rate p that the test will yield a positive result given that the patient has the disease and the false positive rate q that the test will yield a positive result given that the patient does not have the disease. For the test to be of any value, p must exceed q. For if $p = q$, the test does no better than a coin flip to determine whether you have the disease. And if $p < q$, the significance of the test result is simply reversed. What one took to be a positive result is actually a negative result, since it makes it less likely that the patient has the disease, and what one took to be a negative result is actually a positive result, since it makes it more likely that the patient has the disease. The true positive and false positive rates of particular medical tests are often well documented. If we receive a number of independent test reports that are all positive, this does not make us any more confident about the tests in question. Nor would it make us less confident if we were to receive a mixture of positive and negative test reports. Typically we take our own case not to affect the goodness of the test as determined by the false positive and false negative rates.

This is true for medical tests, but it does not hold true in general for partially reliable sources. We may initially distrust a set of sources, but if they independently provide us with the same (or highly coherent) information, our confidence in their reliability is increased. Lewis remarks that if we receive the same item of information from multiple independent witnesses, then our confidence increases that the witnesses are truth-tellers: '[O]n any other hypothesis than that of truth-telling, this agreement [between multiple independent witnesses] is highly unlikely' (1946: 346). The idea of making the reliability of the 'evidentiary mechanism'—say, a witness—endogenous to a model of hypothesis confirmation is also the central idea of what David A. Schum (1988: 258–61) names the 'Scandinavian School of Evidentiary Value'. Within this school, Bengt Hansson (1983: 78), with reference to Martin Edman (1973), imposes the following condition on unreliable witnesses: Conditional on a witness being unreliable, learning the evidence does not teach us anything about the hypothesis. On the other hand, a reliable witness is a truth-teller. We may not know beforehand whether the witness is reliable or not. So we construct a model with witness reliability—thus conceived—as an endogenous variable.

Let *REP* be a binary propositional variable which has as its values the presence or absence of a witness report that constitutes evidence for the hypothesis. Let *REL* be a binary propositional variable which takes as its values the reliability or unreliability of the witness. Similarly, let *HYP* be a binary propositional variable which takes as its values the truth or falsity of the hypothesis. Henceforth, *HYP* will take the place of R_i in our previous discussion, since we will be concerned with the confirmation of a single proposition HYP. Edman's condition can then be expressed as follows:

$$(3.1)\qquad P(\text{HYP}|\neg\text{REL}) = P(\text{HYP}|\text{REP}, \neg\text{REL}).$$

This entails the following identity by the probability calculus (see Appendix C.1):

$$(3.2)\qquad P(\text{REP}|\text{HYP}, \neg\text{REL}) = P(\text{REP}|\neg\text{HYP}, \neg\text{REL}) =: a,$$

where the *randomization parameter a* is defined as the value of these likelihoods. So, we assume that if witnesses are not reliable, then they are like randomizers. It is as if they do not even look at the state of the world to determine whether the hypothesis is true, but rather flip a coin or cast a die to determine whether they will provide a report to the effect that the hypothesis is true. The randomization parameter *a* indicates the chance that they provide a positive report concerning the hypothesis. On the other hand, if the witnesses are reliable, then they say of what is, that it is and of what is not, that it is not:

$$(3.3)\qquad P(\text{REP}|\text{HYP}, \text{REL}) = 1 \text{ and } P(\text{REP}|\neg\text{HYP}, \text{REL}) = 0.$$

Let the *reliability* parameter ρ express the prior probability that a witness is reliable:

$$P(\text{REL}) = \rho. \tag{3.4}$$

Let the parameter h express the prior probability of the hypothesis (which matches the expectance measure a_0 for a single fact variable as defined in Section 1.3):

$$P(\text{HYP}) = h. \tag{3.5}$$

We suppose that the values of a, h, and ρ are contained in the open interval (0, 1)—i.e. randomizers are not stuck on always reporting REP or always reporting ¬REP, the source is neither certainly reliable nor certainly unreliable, and the hypothesis is neither certainly true nor certainly false. Furthermore, we make the rather innocent assumption that *REL* and *HYP* are independent variables:

$$REL \perp\!\!\!\perp HYP. \tag{3.6}$$

This condition states that learning no more than that a witness is reliable (without knowing what report is forthcoming) does not teach us anything about the truth or falsity of the hypothesis.

3.2. ONE WITNESS REPORT

Before we assess Lewis's conjecture that our confidence in the reliability of the witnesses increases when we get the same information from multiple independent witnesses, let us first assess how a single report that the hypothesis is true affects our confidence in the reliability of the witness. We calculate:

$$\begin{aligned} P^*(\text{REL}) &= P(\text{REL}|\text{REP}) \\ &= \frac{P(\text{REL, REP})}{P(\text{REP})} \\ &\quad \text{(by the definition of conditional probability)} \end{aligned} \tag{3.7}$$

$$= \frac{\sum_{HYP} \mathbf{P}(HYP, \text{REL}, \text{REP})}{\sum_{HYP, REL} \mathbf{P}(HYP, REL, \text{REP})} \quad \text{(by expansion)}$$

$$= \frac{\sum_{HYP} \mathbf{P}(\text{REP}|HYP, \text{REL})\ \mathbf{P}(HYP|\text{REL})\ \mathbf{P}(\text{REL})}{\sum_{HYP, REL} \mathbf{P}(\text{REP}|HYP, REL)\ \mathbf{P}(HYP|REL)\ \mathbf{P}(REL)}$$

(by the chain rule)

$$= \frac{\sum_{HYP} \mathbf{P}(\text{REP}|HYP, \text{REL})\ \mathbf{P}(HYP)\ \mathbf{P}(\text{REL})}{\sum_{HYP, REL} \mathbf{P}(\text{REP}|HYP, REL)\ \mathbf{P}(HYP)\ \mathbf{P}(REL)}.$$

(by the independence assumption (3.6))

Table 3.1 presents the sum in the numerator and Table 3.2 presents the sum in the denominator. Hence,

$$\mathbf{P}(\text{REL}|\text{REP}) = \frac{h\rho}{h\rho + a\overline{\rho}} \quad \text{where } \overline{\rho} = 1 - \rho. \tag{3.8}$$

We can now study the relationship between the prior probability **P**(REL) and the posterior probability **P***(REL):

TABLE 3.1 *Calculating the sum in the numerator in (3.7)*

$\mathbf{P}(\text{REP}|\text{REL, HYP})\ \mathbf{P}(\text{REL})\ \mathbf{P}(\text{HYP})$	$1\rho h$
$\mathbf{P}(\text{REP}|\text{REL}, \neg\text{HYP})\ \mathbf{P}(\text{REL})\ \mathbf{P}(\neg\text{HYP})$	$0\rho(1-h)$
	$\sum = h\rho$

TABLE 3.2 *Calculating the sum in the denominator in (3.7)*

$\mathbf{P}(\text{REP}|\text{REL, HYP})\ \mathbf{P}(\text{REL})\ \mathbf{P}(\text{HYP})$	$1\rho h$
$\mathbf{P}(\text{REP}|\text{REL}, \neg\text{HYP})\ \mathbf{P}(\text{REL})\ \mathbf{P}(\neg\text{HYP})$	$0\rho(1-h)$
$\mathbf{P}(\text{REP}|\neg\text{REL, HYP})\ \mathbf{P}(\neg\text{REL})\ \mathbf{P}(\text{HYP})$	$a(1-\rho)h$
$\mathbf{P}(\text{REP}|\neg\text{REL}, \neg\text{HYP})\ \mathbf{P}(\neg\text{REL})\ \mathbf{P}(\neg\text{HYP})$	$a(1-\rho)(1-h)$
	$\sum = h\rho + a(1-\rho)$

(3.9)
$$\Delta_{\text{REL}} = \boldsymbol{P}^*(\text{REL}) - \boldsymbol{P}(\text{REL}) = \frac{h\rho}{h\rho + a\bar{\rho}} - \rho = (h - a)\frac{\rho\bar{\rho}}{h\rho + a\bar{\rho}}.$$

Since the second factor in the expression above is always positive, the sign of Δ_{REL} equals the sign of $h - a$. For instance, for low values of h and for high values of a, $\Delta_{\text{REL}} < 0$ holds. This is to be expected. If we are dubious that the hypothesis is true, then we tend to blame a positive report on the lack of reliability of the witness—and even more so if we know that an unreliable witness would be randomizing with a high probability of providing a positive report. On the other hand, for high values of h and for low values of a, $\Delta_{\text{REL}} > 0$ holds. If we are confident that the hypothesis is true, then we take a positive report to be a reason to increase our trust in the reliability of the witness—and even more so if we know that an unreliable witness would be randomizing with a low probability of providing a positive report. If $a = h$, then $\Delta_{\text{REL}} = 0$ holds—our opinion about the reliability of the witness remains unaffected by her report.

3.3. MULTIPLE WITNESS REPORTS

But now suppose that more and more independent witnesses step forward and provide us with the same report that the hypothesis is true. How does this affect our confidence in the reliability of the witnesses? We will show that the probability that one of the witnesses is reliable conditional on n witnesses all providing positive reports is a monotonically increasing function for $n \geq 1$. Although our confidence in the reliability of the witnesses may drop with the first positive report, once matching reports start coming in, our confidence in the reliability of the witnesses is bound to increase, which confirms Lewis's conjecture.

We are interested in the posterior probability that one of n witnesses is reliable after they have all provided us with positive reports:

(3.10)
$$\boldsymbol{P}^{*(n)}(\text{REL}_i) = \boldsymbol{P}(\text{REL}_i|\text{REP}_1, \ldots, \text{REP}_n).$$

There are two aspects to the stipulation that the witnesses are independent. First, the chance that we will get a positive report from a witness is fully determined by whether that witness is reliable and by whether the hypothesis they report on is true. Learning about other witness reports or about the reliability of other witnesses does not affect this chance:

$$REP_i \perp\!\!\!\perp REP_1, REL_1, \ldots, REP_{i-1}, REL_{i-1}, REP_{i+1}, REL_{i+1}, \ldots, REP_n, REL_n | REL_i, HYP \text{ for } i = 1, \ldots, n. \tag{3.11}$$

Second, the chance that a witness is reliable is independent of the reliability of the other witnesses and of the truth of the hypothesis. Learning about the reliability of other witnesses does not affect the chance that the witness under consideration is reliable:

$$REL_i \perp\!\!\!\perp REL_1, \ldots, REL_{i-1}, REL_{i+1}, \ldots, REL_n, HYP \text{ for } i = 1, \ldots, n. \tag{3.12}$$

To keep the calculations simple, let us define the characteristics of each less than fully reliable witness $i = 1, \ldots, n$ in the same way as we defined a single, less than fully reliable witness earlier:

$$P(\text{REP}_i|\text{HYP}, \neg\text{REL}_i) = P(\text{REP}_i|\neg\text{HYP}, \neg\text{REL}_i) = a, \tag{3.13}$$

$$P(\text{REP}_i|\text{HYP}, \text{REL}_i) = 1 \text{ and } P(\text{REP}_i|\neg\text{HYP}, \text{REL}_i) = 0, \tag{3.14}$$

$$P(\text{REL}_i) = \rho. \tag{3.15}$$

Let's calculate the posterior probability that one of the witnesses is reliable given positive reports from all n witnesses and given equal prior probability that the witnesses are reliable. The procedure is the same as in our earlier calculations. We apply the definition of conditional probability, expand both numerator and denominator, apply the chain rule, work out the independences in (3.11) and (3.12) and insert the values in (3.5) and (3.13) through (3.15).[1] This yields:

[1] The proof is a special case of the more general proof of (5.2), which can be found in Appendix E.2.

(3.16) $$\mathbf{P}^{*(n)}(\mathrm{REL_i}) = \mathbf{P}(\mathrm{REL_i}|\mathrm{REP_1}, \ldots, \mathrm{REP_n}) = \frac{h\bar{x}}{h + \bar{h}x^n}.$$

$x \in (0, 1)$ represents the likelihood ratio of a single report:

(3.17) $$x = \frac{\mathbf{P}(\mathrm{REP_i}|\neg\mathrm{HYP})}{\mathbf{P}(\mathrm{REP_i}|\mathrm{HYP})} = \frac{a\bar{\rho}}{\rho + a\bar{\rho}}.$$

$\mathbf{P}^{*(n)}(\mathrm{REL_i})$ is an increasing function of n for $n \geq 1$, which has the following limit as n goes to infinity:

(3.18) $$\lim_{n \to \infty} \mathbf{P}^{*(n)}(\mathrm{REL_i}) = \bar{x} = \frac{\rho}{\rho + a\bar{\rho}}.$$

This limit is smaller than 1 for a greater than 0. Why do we reach this ceiling in the limit? Should our confidence that the witnesses are reliable not converge to certainty as the number of positive reports increases? Let's see why this is not the case.

Intuitively, as the number of witnesses providing positive reports goes to infinity, our degree of confidence that the hypothesis is true converges to 1. Indeed,[2]

(3.19) $$\mathbf{P}^{*(n)}(\mathrm{HYP}) = \mathbf{P}(\mathrm{HYP}|\mathrm{REP_1}, \ldots, \mathrm{REP_n}) = \frac{h}{h + \bar{h}x^n},$$

so that

(3.20) $$\lim_{n \to \infty} \mathbf{P}^{*(n)}(\mathrm{HYP}) = 1.$$

But even though our degree of confidence that the hypothesis is true converges to 1 as n goes to infinity, a positive report may still have come from a randomizing unreliable witness. To see this, suppose that we were indeed certain that the hypothesis is true. Let $\mathbf{P}^{\#}$ express

[2] The proof is a special case of the more general proof of (5.1), which can be found in Appendix E.1.

the probability distribution given this background knowledge. Note that $P^{\#}(\text{REL}_i) = \rho$ by (3.12) and (3.15), that $P^{\#}(\text{REP}_i|\text{REL}_i) = 1$, since a reliable witness will provide a positive report about a true hypothesis, and that $P^{\#}(\text{REP}_i|\neg\text{REL}_i) = a$, since there is a chance a that an unreliable witness will provide a positive report. On the assumption that we know that the hypothesis is true, the chance that a witness is reliable given that she provided a positive report equals

$$(3.21) \qquad P^{\#}(\text{REL}_i|\text{REP}_i) = \frac{P^{\#}(\text{REP}_i|\text{REL}_i)\, P^{\#}(\text{REL}_i)}{\sum_{REL_i} P^{\#}(\text{REP}_i|REL_i)\, P^{\#}(REL_i)} = \frac{\rho}{\rho + a\overline{\rho}} = \overline{x},$$

which matches the result in (3.18).

3.4. AN UPPER LIMIT FOR RELIABILITY?

Lewis claims that when we receive lots of witness reports pointing in the same direction, any hypothesis other than that they are telling the truth is *highly unlikely*. And indeed, we do seem to put our full trust in witnesses who independently provide us with the same reports. This seems to suggest that $\lim_{n\to\infty} P^{*(n)}(\text{REL}_i)$ should converge to1 and not to some upper limit lower than 1, as we argued in Section 3.3. We will show two distinct ways to account for Lewis's intuition.

The first way is suggested by Lewis himself. Lewis accounts for his claim by pointing out that 'the story any one false witness might tell [is] one out of so very large a number of equally possible choices' (1946: 346). Lewis envisions that we are not dealing with some question to which the answer is either yes or no, but rather with a question that may allow for a great many possible answers. The chance that an unreliable witness would pick one particular answer, viz. the one that was provided by the other witnesses, out of these many possible answers is indeed negligible. What Lewis's suggestion amounts to in our terminology is that we should let the randomization parameter a converge to 0. And indeed,

$$\lim_{a \to 0} \frac{\rho}{\rho + a\bar{\rho}} = 1. \quad (3.22)$$

The other way comes from relaxing the independence assumption in (3.12). This assumption is indeed very strong. We are asked to believe the following. Suppose that we assign a certain prior probability ρ that a particular source is reliable. Now we come to learn that all other sources are reliable or we come to learn that all other sources are unreliable. Then the independence assumption requires that we not revise the chance ρ that this one particular source is reliable. This is highly unusual. If we come to learn that even a small range of oracles (tea leaves, coffee grounds, ravens, etc.) are unreliable, then we turn very sceptical about other kinds of oracles. Or if we come to learn that some techniques of medical diagnosis in Borneo indeed do have diagnostic value, then we are more willing to lend an ear to what other such techniques have to offer. How does relaxing the independence assumption in (3.12) affect the question at hand? Does it affect the upper limit that our degree of confidence in the reliability of the witnesses reaches when we receive multiple positive reports?

To investigate this question, we construct a model in which we stipulate that either all of the witnesses are fully reliable or all of the witnesses are fully unreliable. In this case, the assignment of the reliability to the witnesses exhibits an extreme form of positive relevance. This assumption is, of course, just as unrealistic as the independence assumption that we stipulated before. The realistic cases are to be found in the range between extreme positive relevance and independence. If we study what happens in these extreme cases then we will also come to understand how our degree of confidence behaves in the range of more realistic intermediate cases.

Let us define the variable *REL* which can take on two values, REL—that all witnesses are reliable—and ¬REL—that none of the witnesses are reliable. As before, we assign a randomization parameter a for unreliable witnesses, and we assume that a reliable witness says of what is, that it is, and of what is not, that it is not; and we also assign a prior probability that all the witnesses will be reliable:

$$P(\text{REP}_i|\text{HYP}, \neg\text{REL}) = P(\text{REP}_i|\neg\text{HYP}, \neg\text{REL}) = a, \quad (3.23)$$

$$P(\text{REP}_i|\text{HYP, REL}) = 1 \text{ and } P(\text{REP}_i|\neg\text{HYP, REL}) = 0, \tag{3.24}$$

$$P(\text{REL}) = \rho. \tag{3.25}$$

Furthermore, there are two innocent independence assumptions which do not affect the aspect of independence that we are doing away with in our new model. First, in the absence of any knowledge about witness reports, learning that the witnesses are reliable does not teach us anything about whether the hypothesis is true:

$$REL \perp\!\!\!\perp HYP. \tag{3.26}$$

Second, given that the hypothesis is true (or not) and given that the witnesses are all reliable (or all unreliable), learning about some witness reports does not teach us anything about others:

$$REP_i \perp\!\!\!\perp REP_1, \ldots, REP_{i-1}, REP_{i+1}, \ldots, REP_n|REL, HYP. \tag{3.27}$$

This assumption implies that unreliable witnesses are truly randomizing. They do not just repeat what others say, but it is effectively as if they flip a coin or cast a die to determine what they are going to say.

Let P' be the probability distribution over $REL, REP_1, \ldots, REP_n, HYP$ which satisfies our assumptions in (3.23) through (3.27). In Appendix C.2, we calculate the posterior probability that a witness is—and hence all witnesses are—reliable after we have received positive reports from all n witnesses:

$$\begin{aligned} P'^{*(n)}(\text{REL}) &= P'(\text{REL}|\text{REP}_1, \ldots, \text{REP}_n) \\ &= \frac{h\rho}{h\rho + a^n\overline{\rho}}. \end{aligned} \tag{3.28}$$

Clearly, as n goes to infinity, the posterior probability that the witnesses are all reliable converges to 1.

$$\lim_{n\to\infty} P'^{*(n)}(\text{REL}) = 1. \tag{3.29}$$

This is easy to understand. As we get more and more positive reports and nothing but positive reports, the chance that we received such reports from randomizing witnesses becomes negligible. And since either all the witnesses are reliable or all the witnesses are unreliable (and hence randomizers), the chance that the witnesses are reliable converges to certainty.

Our results allow for a qualified endorsement of Lewis's claim within the framework of the evidentiary theory of value. Whether a single report increases or decreases our confidence in the reliability of the witness is contingent on the prior probability of the hypothesis and on the value of the randomization parameter. An unlikely report from a witness who, if unreliable, is likely to present us with a positive report, decreases our confidence in the reliability of the witness. A likely report from a witness who, if unreliable, is unlikely to present us with a positive report, increases our confidence in the reliability of the witness. But once the first positive report is in, subsequent positive reports from independent witnesses progressively increase our confidence in the witnesses, no matter what the prior probability of the hypothesis and the value of the randomization parameter are. However, the posterior probability that a particular witness is reliable will reach an upper limit that is a function of the randomization parameter and the reliability parameter.

To avoid this result, we can either let the randomization parameter converge to 0, by arguing that it would be very unlikely that someone would hit on precisely the same story as the other witnesses by chance. Or, we can relax a particular independence assumption, viz. learning that some witnesses are reliable teaches us nothing about the reliability of the other witnesses. To relax this assumption, we examined how the posterior probability of the reliability of the witnesses behaves when we make the assumption on the other extreme of the spectrum, viz. that one witness is reliable if and only if all witnesses are reliable. In this case, the chance that the witnesses are reliable converges to 1 as the number of witnesses goes to infinity. In reality, as we continue to learn that more and more witnesses are reliable, we become more confident that the next witness will be reliable as well. But there are many ways of walking the route from independence to extreme positive relevance. We turn to the challenge of spelling out an intermediate position in the next section.

3.5. BAYESIAN NETWORKS

Over the last two decades the theory of Bayesian Networks has been erected in artificial intelligence on the dual pillars of graph theory and the theory of conditional independence structures. Although the theory of Bayesian Networks will not be essential to our results, it is useful for providing graphical representations of conditional independencies between variables, and for easily determining the closure of a set of conditional dependencies. We will retell part of the story that we have told so far in terms of Bayesian Networks. We do so to show how much easier life can be with a tiny bit of theory. For simple models, Bayesian Networks will seem gratuitous. But as the conditional independence structures become more complex, Bayesian Networks become a welcome (though not an indispensable) tool. Spelling out intermediate positions between extreme positive relevance and independence will certainly be easier with the aid of Bayesian Networks.[3] Furthermore, in the next two chapters we will find it helpful to represent sets of conditional independences in terms of Bayesian Networks.

In the first two chapters, the reliability of the instrument was defined exogenously. There was a certain chance p that the instrument will indicate that a fact obtains (in other words, provide a positive report) given that the fact obtains and a certain chance q that the instrument will indicate that a fact obtains given that the fact does not obtain. These chances were supposed to remain fixed, no matter how many positive reports we receive. Let us see how to represent this very simple model in a Bayesian Network.

A Bayesian Network organizes the variables into a *Directed Acyclical Graph* (DAG), which encodes a set of (conditional) independences. A DAG is a set of nodes and a set of arrows between some of the nodes. The only constraint is that one cannot run into a cycle by following the arrows. Each node represents a propositional variable, which can take any number of mutually exclusive and exhaustive values. Consider two nodes linked by an arrow. The node at the tail is the *parent node* of the node at the head and the node at the head is the *child node* of the node at the tail. The arrows in the network have a precise probabilistic meaning that we will spell out below. For now, we will just say that the following heuristic governs the construction of a

[3] *Cf.* Bovens and Olsson (2000) and Bovens and Hartmann (2002).

DAG: There is an arrow between two nodes if the variable in the parent node has a direct influence on the variable in the child node. Furthermore, *root nodes* are unparented nodes and a node at a tail is a *descendant node* of a node at a head if the former is a child node of the latter, or a child node of a child node of the latter etc.

The variable *HYP* can take two values—HYP, that is, that the hypothesis is true, and ¬HYP, that is, that the hypothesis is false. The variable *REP* can also take two values—REP, that is, that there is a report to the effect that the hypothesis is true, and ¬REP, that is, that there is no report to the effect that the hypothesis is true.[4] When the reliability of the instrument is exogenous to the model, the variable *REP* is directly influenced by the variable *HYP*. When the reliability of the instrument is endogenous to the model, we specify a variable *REL* that can take on the value REL, that is, that the instrument is reliable, and the value ¬REL, that is, that the instrument is not reliable. Whether there is a report to the effect that the hypothesis holds is directly influenced by, and only by, whether the test consequence holds or not and whether the instrument is reliable or not.

We construct the graph in Figure 3.1 for the case in which the reliability of the witness is defined exogenously. *HYP* is the variable in the parent node and *REP* is the variable in the child node. We construct the graph in Figure 3.2 for the case in which the reliability of the witness is defined endogenously. *HYP* and *REL* are the variables in the parent nodes and *REP* is the variable in the child node.

To turn a DAG into a Bayesian Network one more step is required. We need the marginal probability distributions for the variables in the root nodes of the graph, and we need the conditional probability distributions for the variables in the child nodes, given any combination of values of the variables in their respective parent nodes. We have assigned these values in Figures 3.1 and 3.2 in accordance with the parameters that we specified earlier.

The arrows in a Bayesian Network have a precise probabilistic meaning: They carry information about the independence relations between the variables in the Bayesian Network. This information is expressed by the *Parental Markov Condition*:[5]

[4] One might also assume that the information source has to give a definite answer, as we will do in the next chapter. Then ¬REP is the proposition that there is a report that the hypothesis is false.

[5] Pearl (2000: 19). Spirtes *et al.* (2000: 11) call this condition the *Markov Condition*.

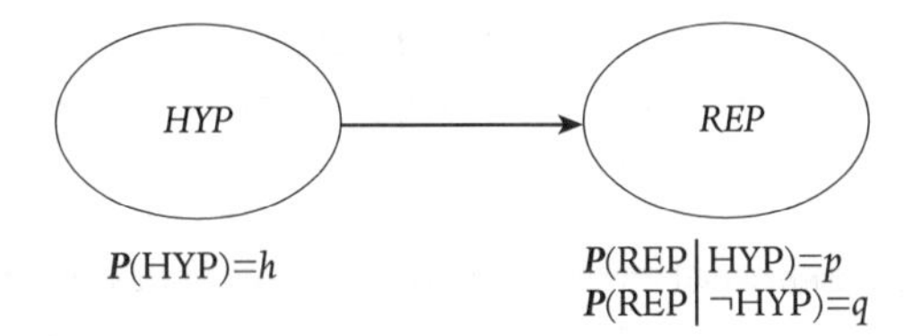

FIG. 3.1 The Bayesian Network for a single report with reliability defined exogenously

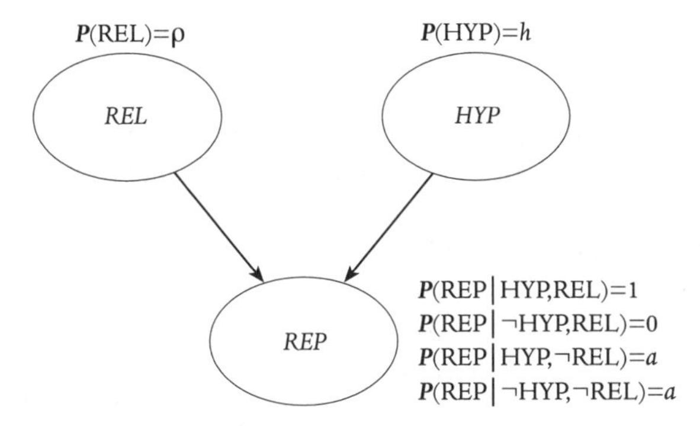

FIG. 3.2 The Bayesian Network for a single report with reliability defined endogenously

(PMC) A variable represented by a node in the Bayesian Network is independent of all variables represented by its non-descendent nodes in the Bayesian Network, conditional on all variables represented by its parent nodes.

In the Bayesian Network in Figure 3.1 there are no (conditional) independences. The Bayesian Network in Figure 3.2 is constructed based on the following assumption of conditional independence:

$$REL \perp\!\!\!\perp HYP. \tag{3.30}$$

This is precisely the independence that we stipulated in (3.6).

What's so great about Bayesian Networks? A Bayesian Network contains information about the independence relations between the variables, a marginal probability distribution for each variable in a root node, and a conditional probability distribution for each variable in a child node given any combination of values of the variables of its parent nodes. A central theorem in the theory of Bayesian Networks states that a joint probability distribution over any combination of variables in the network is equal to the product of the marginal probabilities and conditional probabilities for these values as expressed in the network (Neapolitan 1990: 162–4). For example, suppose we are interested in the joint probability of HYP, REP, and ¬REL. We can read the joint probability directly off Figure 3.2:

$$
\begin{aligned}
(3.31)\quad \boldsymbol{P}(\text{HYP}, \neg\text{REL}, \text{REP}) &= \boldsymbol{P}(\text{HYP})\,\boldsymbol{P}(\neg\text{REL})\,\boldsymbol{P}(\text{REP}|\text{HYP}, \neg\text{REL}) \\
&= ah\bar{\rho}.
\end{aligned}
$$

We can now calculate, say, the posterior probability that the hypothesis is true after we have received a report to this effect:

$$
\begin{aligned}
(3.32)\quad \boldsymbol{P}^*(\text{HYP}) &= \boldsymbol{P}(\text{HYP}|\text{REP}) \\
&= \boldsymbol{P}\frac{(\text{HYP}, \text{REP})}{\boldsymbol{P}(\text{REP})} \\
&= \frac{\sum_{REL} \boldsymbol{P}(\text{HYP}, REL, \text{REP})}{\sum_{HYP,\, REL} \boldsymbol{P}(HYP, REL, \text{REP})} \\
&= \frac{\sum_{REL} \boldsymbol{P}(\text{HYP})\,\boldsymbol{P}(REL)\,\boldsymbol{P}(\text{REP}|\text{HYP}, REL)}{\sum_{HYP,\, REL} \boldsymbol{P}(HYP)\,\boldsymbol{P}(REL)\,\boldsymbol{P}(\text{REP}|HYP, REL)} \\
&= \frac{h(\rho + a\bar{\rho})}{h(\rho + a\bar{\rho}) + \bar{h}a\bar{\rho}} \\
&= \frac{h}{h + \bar{h}x} \quad \text{with } x \text{ as defined in (3.17).}
\end{aligned}
$$

In our model in Chapters 1 and 2 we envisioned a situation in which multiple independent witnesses provide us with different reports. We defined an independent witness as a witness who is strictly influenced by the fact that she reports on and not by other facts or by other witnesses. Now suppose that there is a particular disease that is characterized by

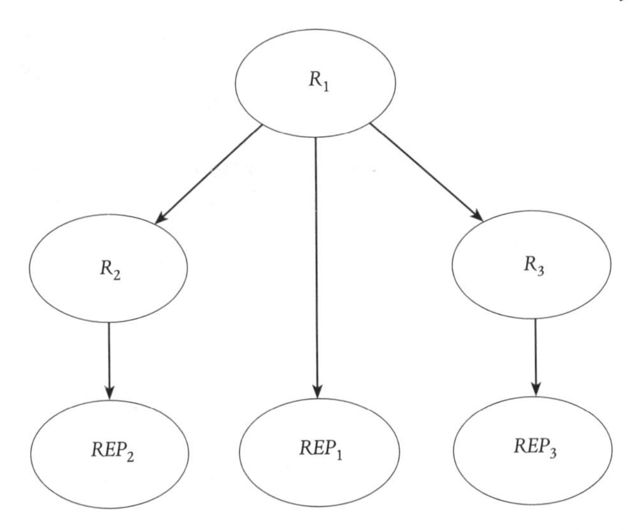

Fig. 3.3 The Bayesian Network for reports about multiple facts with independent witnesses and reliability defined exogenously

(R_1) the presence of certain bacteria in the blood stream. There are two independent symptoms that typically accompany this disease, viz. (R_2) internal bleeding and (R_3) an inflammation of the lymph system. We receive less than fully reliable reports from independent sources that (REP_1) the bacteria are present in someone's blood stream, that (REP_2) internal bleeding has occurred and that (REP_3) her lymph nodes are inflamed.

Various conditional independence relations can hold between the fact variables, that is, the variables whose positive values are the content of the reports. A standard case of independent evidence for a hypothesis is that of independent symptoms of a disease. The disease variable R_1 has a direct influence on each of the symptoms R_2 and R_3, but neither one of the symptoms have a direct influence on each other. Hence we draw an arrow from R_1 to both R_2 and R_3 in Figure 3.3. By the Parental Markov Condition we read off the following independence:

$$R_2 \perp\!\!\!\perp R_3 | R_1. \tag{3.33}$$

Let the prior probability distribution over the root variable R_1 reflect the frequency of the disease in the relevant population and let the conditional probabilities $\boldsymbol{P}(\mathrm{R_i}|\mathrm{R_1})$ and $\boldsymbol{P}(\mathrm{R_i}|\neg\mathrm{R_1})$ reflect the strength of the connection between the disease and the respective symptoms for $i = 2, 3$. This conditional independence relation between the fact variables and this probabilistic information is merely for the purpose of illustration. What is really essential to this model is that the report variables REP_i are directly influenced by the respective fact variables R_i, and not by other fact variables, nor by other report variables. So we draw arrows from each R_i to each REP_i for $i = 1, 2, 3$ respectively. By the Parental Markov Condition we can read off the conditional independences from the graph in Figure 3.3:

$$(3.34) \qquad REP_i \perp\!\!\!\perp REP_1, \ldots, REP_{i-1}, REP_{i+1}, \ldots, REP_n, R_1, \ldots, R_{i-1}, R_{i+1}, \ldots, R_n | R_i.$$

This is precisely the independence condition that we laid out in Chapter 1. Using the methodology that we outlined earlier it is easy to see how we can calculate

$$(3.35) \qquad \mathbf{P}(R_1, \ldots, R_n, REP_1, \ldots, REP_n) = \mathbf{P}(REP_1|R_1) \times \ldots \times \mathbf{P}(REP_n|R_n)\ \mathbf{P}(R_1, \ldots, R_n)$$

for particular values of the variables $R_1, \ldots, R_n, REP_1, \ldots, REP_n$. In the case at hand, we can go one step further because of the conditional independence between the fact variables in (3.33):

$$(3.36) \qquad \mathbf{P}(R_1, R_2, R_3, REP_1, REP_2, REP_3) = \mathbf{P}(REP_1|R_1)\ \mathbf{P}(REP_2|R_2)\ \mathbf{P}(REP_3|R_3)\ \mathbf{P}(R_3|R_1)\ \mathbf{P}(R_2|R_1)\ \mathbf{P}(R_1).$$

These joint probabilities can then be used in calculating conditional probabilities and marginal probabilities according to the standard rules of the probability calculus.

If we wish to define the reliability of the witnesses endogenously, then we can attach report variables that are parented by both fact variables and reliability variables as in Figure 3.4. Remember our earlier result in (1.5) of Chapter 1 in which the reliability of the witnesses is exogenous to the model, with $\bar{r} = q/p$:

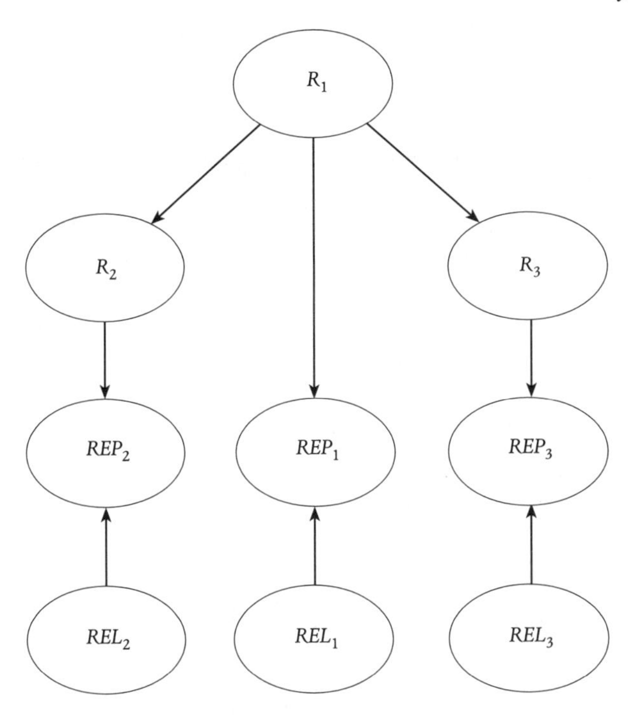

Fig. 3.4 The Bayesian Network for reports about multiple facts with independent witnesses and reliability defined endogenously

$$P^*(\mathrm{R}_1, \ldots, \mathrm{R}_n) = \frac{a_0}{\sum_{i=0}^{n} a_i (q/p)^i} = \frac{a_0}{\sum_{i=0}^{n} a_i \bar{r}^i}. \tag{3.37}$$

In Figure 3.4, R_i is a non-descendant of REL_i, and R_i is independent of REL_i given REL_i's parents, i.e. given the null set. By the Parental Markov Condition:[6]

[6] Strictly speaking, the Parental Markov condition permits us to read off $REL_i \perp\!\!\!\perp R_1, \ldots, R_n, REP_1, \ldots, REP_{i-1}, REP_{i+1}, \ldots, REP_n$ which entails (3.38) by the Axiom of Decomposition. A Bayesian Network contains the closure of the conditional

$$(3.38) \qquad REL_i \perp\!\!\!\perp R_i \text{ for } i = 1, \ldots, n.$$

We can now calculate the value of p:

$$(3.39) \quad p = \mathbf{P}(\mathrm{REP_i}|\mathrm{R_i}) = \sum_{REL_i} \mathbf{P}(\mathrm{REP_i}, REL_i|\mathrm{R_i})$$

$$= \sum_{REL_i} \mathbf{P}(\mathrm{REP_i}|REL_i, \mathrm{R_i})\mathbf{P}(REL_i|\mathrm{R_i})$$

$$= \sum_{REL_i} \mathbf{P}(\mathrm{REP_i}|REL_i, \mathrm{R_i})\mathbf{P}(REL_i) \quad \text{(by (3.38))}$$

$$= \rho + a\bar{\rho}.$$

And similarly for q:

$$(3.40) \qquad q = \mathbf{P}(\mathrm{REP_i}|\neg\mathrm{R_i}) = a\bar{\rho}$$

for $i = 1, \ldots, n$. We substitute these expressions into (3.37):[7]

$$(3.41) \qquad \mathbf{P}^*(\mathrm{R_1}, \ldots, \mathrm{R_n}) = \frac{a_0}{\sum_{i=0}^{n} a_i \left(\frac{a\bar{\rho}}{\rho + a\bar{\rho}}\right)^i},$$

which allows us to calculate directly the posterior probability that the information is true for the model in which the reliability of the witnesses is defined endogenously.

Other variations can be introduced into a model of independent and relatively unreliable witnesses as well. For instance, we could suppose that an unreliable witness is somewhat more likely to tell the truth than to lie, i.e. $\mathbf{P}(\mathrm{REP}|\mathrm{HYP}, \mathrm{REL}) := a^+ > a^- =: \mathbf{P}(\mathrm{REP}|\neg\mathrm{HYP}, \mathrm{REL})$. Or we could assume that we are ignorant about the chance that the witnesses are reliable and represent this ignorance by some density function over

independences that can be read off by the Parental Markov Condition under the semi-graphoid axioms, one of which is the Axiom of Decomposition. The semi-graphoid axioms are presented in Pearl (1997: 82–90). They first occur in Dawid (1979) and Spohn (1980). The standard way of reading independences off a Bayesian Network is by means of the d-separation criterion. For details on the d-separation criterion, see Pearl (1997: 117–18), Neapolitan (1990: 202–7) and Jensen (1996: 12–14).

[7] Remember that a_i are the components of the weight vector $< a_0, \ldots, a_n >$ as defined in Chapter 1 and are entirely distinct from the randomization parameter a.

$\rho \in (0, 1)$. Alternatively, we could assume that we are ignorant about the chance that an unreliable witness will provide a positive report and represent this ignorance by some density function over $a \in (0, 1)$.

As long as the witnesses remain independent and relatively unreliable, one can calculate the values of p and q in any of these more complex models and substitute these expressions into formula (3.37). Our argument in Chapter 1 to the effect that there cannot exist a coherence ordering and the construction of a quasi-ordering coherence in Chapter 2 only rests on the assumption that the witnesses are independent and relatively unreliable. These results were obtained by means of a simple model in which witness reliability is defined exogenously, and they hinge on (3.37). But they could also have been obtained from more complex models in which witness reliability is defined endogenously, since the posterior probability that all the witness reports are true in these models can be expressed in terms of (3.37). These complexities are incidental to our results in Chapters 1 and 2.

Let us now turn to a Bayesian Network presentation of the case in which n independent witnesses provide us with the same report. This is accomplished by contracting all of the fact variable nodes into a single node with the fact variable *HYP* whose positive value is the content of the report in question. The Bayesian Network is presented in Figure 3.5. Note that the independences which we stipulated before in (3.11) and (3.12) can be read off by means of the Parental Markov Condition. If we wish to substitute extreme positive relevance for independence we contract all of the nodes with the reliability variables REL_i for witnesses $i = 1, \ldots, n$ into a single node with the variable REL, as is shown in Figure 3.6. REL takes on the values REL, i.e. that all the witnesses are reliable, and $\neg$REL, i.e. that all of the witnesses are unreliable. Again, we can read off the independences in (3.26) and (3.27).

As we mentioned earlier, what is typically the case is that some intermediate degree of positive relevance holds between the reliability of the witnesses. A Bayesian Network comes in handy if one wishes to model this situation. We introduce an additional binary variable *SR* (for *Super-Reliability*) into the model. We draw arrows from *SR* to the various REL_is, specify a prior probability $\mathbf{P}(\text{SR}) = u$ and conditional probabilities $1 > \mathbf{P}(\text{REL}_i|\text{SR}) =: s > t := \mathbf{P}(\text{REL}_i|\neg\text{SR}) > 0$. The Bayesian Network is represented in Figure 3.7. If we were to set $s = t$, then

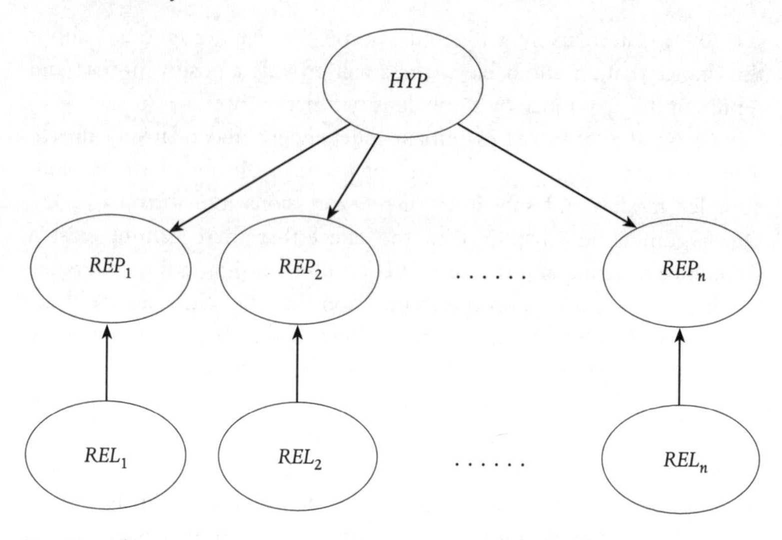

Fig. 3.5 The Bayesian Network for multiple reports about a single fact with independent witnesses and reliability defined endogenously

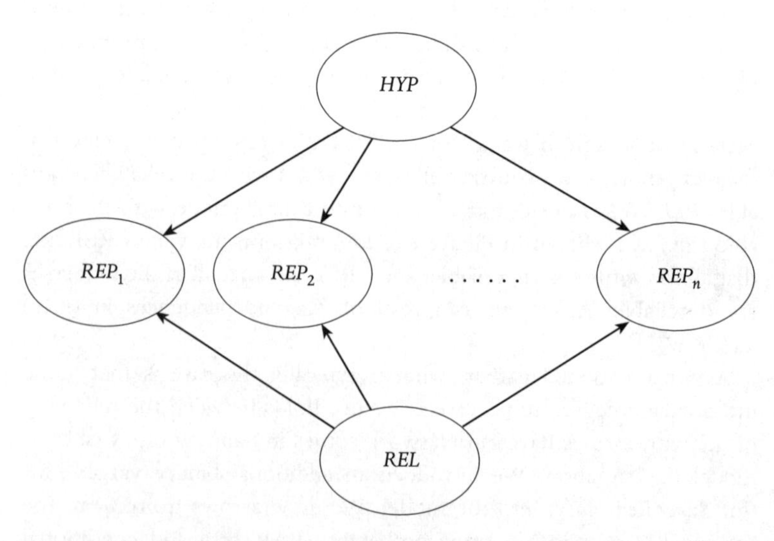

Fig. 3.6 The Bayesian Network for multiple reports about a single fact with reliability defined endogenously and with extreme positive relevance in the specification of witness reliability

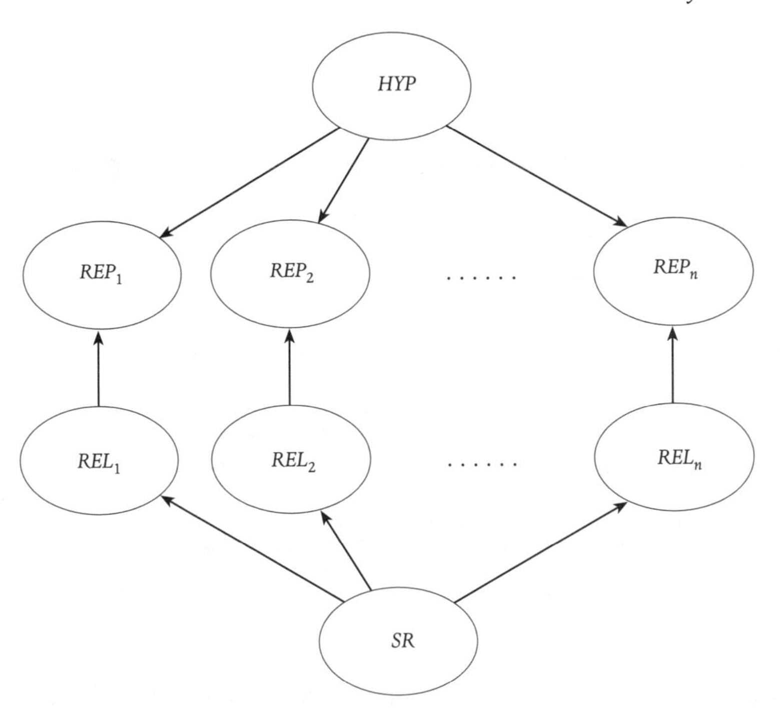

FIG. 3.7 The Bayesian Network for multiple reports about a single fact with reliability defined endogenously and permitting different degrees of dependence in the specification of witnesses reliability

no matter what the value of u, learning that $n - 1$ witnesses are reliable won't teach us anything about the reliability of the nth witness. Hence, we would be back to the model with independence in Figure 3.5. If we were to set $s = 1$ and $t = 0$, then either all the witnesses are reliable with probability u or they are all unreliable with probability $\overline{u}$. Hence we would be back to the model with extreme positive relevance in Figure 3.6. In Appendix C.3, we calculate the posterior probability that the nth witness is reliable given that $n - 1$ witnesses are reliable:

$$(3.42) \qquad P^{*(n-1)}(\text{REL}_n) = P(\text{REL}_n | \text{REL}_1, \ldots, \text{REL}_{n-1}) = \frac{us^n + \overline{u}t^n}{us^{n-1} + \overline{u}t^{n-1}}.$$

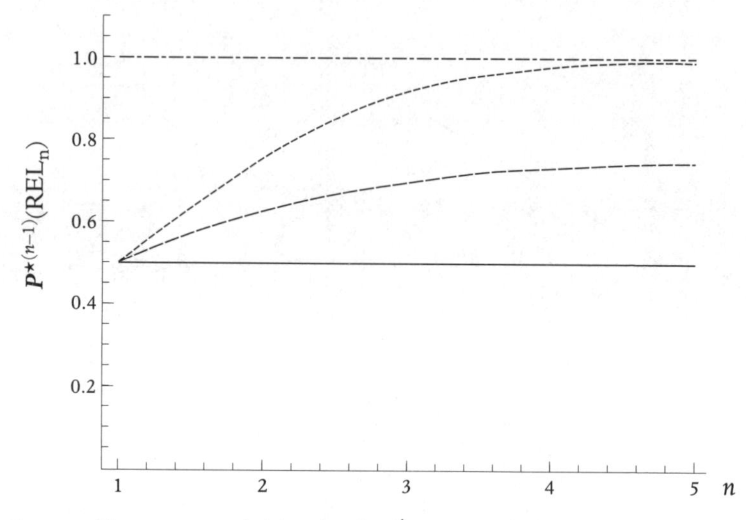

FIG. 3.8 The posterior probability that the nth witness is reliable after learning that all $n-1$ witnesses are reliable for extreme positive relevance (dot-dashed line), for strong positive relevance (dotted line), for weak positive relevance (dashed line) and for independence (full line)

In Figure 3.8, we plot the posterior probability that the nth witness is reliable. We introduced a set of values for independence, a set of values for extreme positive relevance and two sets of values for intermediate cases. To permit comparisons, we chose the values so that the prior probability that a witness is reliable always equals $P(\text{REL}_i) = \text{su} + \text{t}\bar{\text{u}} = .50$ for $i = 1, \ldots, n$. For the case of independence we set the values at $s = .50$ and $t = .50$—the value of u is of no consequence in this case. For these values, $P^{*(n-1)}(\text{REL}_n)$ as a function of $n-1$ is constant at .50, since learning that $n-1$ witnesses are reliable teaches us nothing about the reliability of the nth witness. For extreme positive relevance we set the values at $s = 1, t = 0$ and $u = .5$. Then $P^{*(n-1)}(\text{REL}_n)$ as a function of $n-1$ is constant at 1, since learning that $n-1$ witnesses are reliable teaches us that the nth witness must be reliable as well, even for $n = 2$. We also plotted two intermediate cases between these extremes. There is a stronger case ($s = 1$, $t = 1/4$ and $u = 1/3$) and a weaker case of positive relevance ($s = 3/4$, $t = 1/4$ and $u = 1/2$). In the stronger case the posterior probability $P^{*(n-1)}(\text{REL}_n)$ increases more rapidly as a function of $n-1$ than in the weaker case.

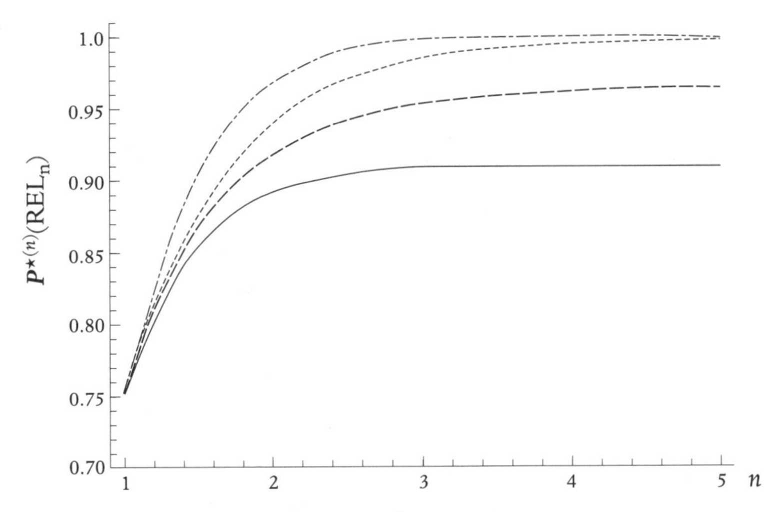

Fig. 3.9 The posterior probability that the nth witness is reliable after receiving n concurring reports that the hypothesis is true for extreme positive relevance (dot-dashed line), for strong positive relevance (dotted line), for weak positive relevance (dashed line) and for independence (full line) for $a = .1$ and $h = .3$

Next, we evaluate how the posterior probability of the reliability of the nth witness increases as more and more witness reports come in (see Appendix C.4 for the derivation):

$$(3.43)\quad P^{*(n)}(\mathrm{REL_n}) = P(\mathrm{REL_n}|\mathrm{REP_1}, \ldots, \mathrm{REP_n}) = \frac{h[us(s+a\bar{s})^{n-1} + \bar{u}t(t+a\bar{t})^{n-1}]}{h[u(s+a\bar{s})^n + \bar{u}(t+a\bar{t})^n] + \bar{h}[u(a\bar{s})^n + \bar{u}(a\bar{t})^n]}.$$

Figure 3.9 plots this function for increasing n with the randomization parameter set at $a = .1$ and the prior probability of the hypothesis set at $h = .3$. With a low randomization parameter, the witness reliability increases to .75 after a single report, and it continues to rise after more positive reports come in. The strongest increase is present in the case of extreme positive relevance, the weakest in the case of independence. As we have pointed out earlier, the function attains a lower plateau for independent witnesses, because however confident we may be that a particular positive report is true after receiving numerous similar reports, independence forces us to entertain the possibility that the

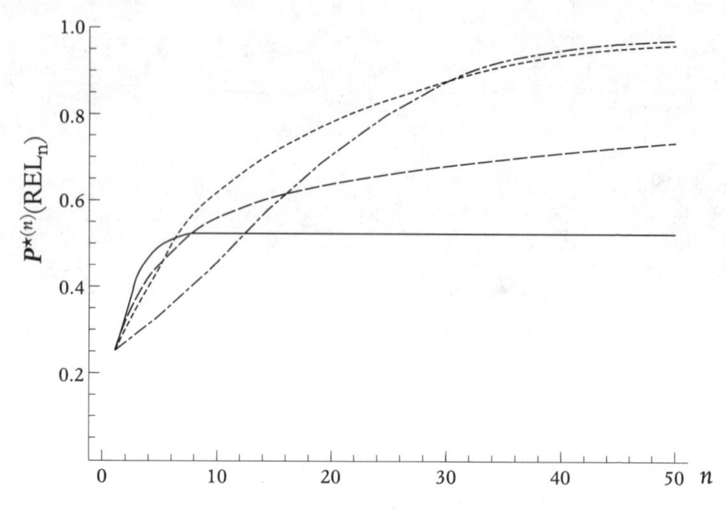

Fig. 3.10 The posterior probability that the nth witness is reliable after receiving n concurring reports that the hypothesis is true for extreme positive relevance (dot-dashed line), for strong positive relevance (dotted line), for weak positive relevance (dashed line) and for independence (full line) for $a = .9$ and $h = .3$

positive report came from an unreliable witness who happened to hit on the right answer by accident.

However, if the randomization parameter is high, then the first positive reports make us suspicious about the reliability of the witnesses. For example, when $a = .9$ and $h = .3$, the witness reliability drops to .25 after a single report, and then slowly increases as more and more positive reports come in. The suspicion lingers the longest for extreme positive relevance, while the function rises faster as we move from extreme positive relevance towards independence. Figure 3.10 illustrates this. So, in a case of positive relevance our confidence in the reliability of the witnesses after a small number of concurring witness reports may be lower than in a case of independence, ceteris paribus, but this trend will reverse as more such reports come in.[8]

[8] Whether the curves for extreme positive relevance and for independence cross is determined by the values of both a and h. It is possible to determine for which values of a and h this crossing occurs; let it suffice to say here that a necessary condition for this crossing is that $a > h$.

3.6 JURY VOTING

We will now consider an application of our model of witnesses with intermediate positive relevance for witness reliability. Christian List (2003) has derived a formula that represents the chance that a suspect is guilty, given a particular percentage of guilty votes by the jury. Here is how it works. Say there are n_+ jurors who deem the suspect guilty and n_- voters who deem the suspect innocent. We assign a prior probability h to the proposition HYP, i.e. that the suspect is guilty. Furthermore, we ascribe a certain competency to the jurors. This competency is the chance that a juror detects the truth. We assume that the chance that a juror finds the suspect innocent, given that the suspect is guilty, is the same as the chance that she finds the suspect guilty, given that the suspect is innocent:

$$\textbf{P}(\neg\text{REP}|\text{HYP}) = 1 - p = q = \textbf{P}(\text{REP}|\neg\text{HYP}). \tag{3.44}$$

We calculate the posterior probability that a suspect is guilty when n_+ voters deem the suspect guilty and n_- voters deem the suspect innocent. Let the total number of votes be $n := n_+ + n_-$. List shows that this chance is only a function of h, p and the *absolute margin* $\delta := n_+ - n_-$. (The same result is proven in Hawthorne (1996) within a more general framework.)

List's model corresponds to the simple Bayesian Network in Figure 3.11. By the Parental Markov Condition the following conditional independence holds:

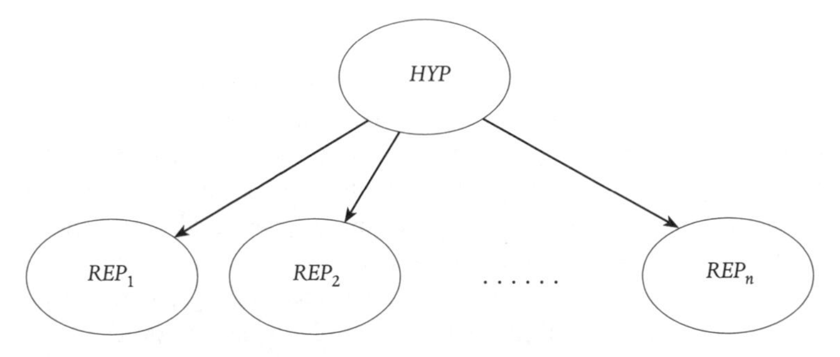

Fig. 3.11 A Bayesian Network for a vote on the guilt or innocence of a suspect by independent voters

$$(3.45)\qquad REP_i \perp\!\!\!\perp REP_1, \ldots, REP_{i-1}, REP_{i+1}, \ldots, REP_n | HYP.$$

We calculate the posterior probability of the hypothesis:

$$(3.46)\quad \mathbf{P}(\mathrm{HYP}|n_+\mathrm{REPs}, n_-\neg\mathrm{REPs}) = \frac{h}{h + \bar{h}x}$$

$$\text{with } x = \frac{\mathbf{P}(n_+\mathrm{REPs}, n_-\neg\mathrm{REPs}|\neg\mathrm{HYP})}{\mathbf{P}(n_+\mathrm{REPs}, n_-\neg\mathrm{REPs}|\mathrm{HYP})} \text{ (by Bayes Theorem)}$$

$$= \frac{q^{n_+}\bar{q}^{n_-}}{p^{n_+}\bar{p}^{n_-}} \text{(by (3.45))}$$

$$= \frac{\bar{p}^{n_+}p^{n_-}}{p^{n_+}\bar{p}^{n_-}} \text{(by (3.44))}$$

$$= \left(\frac{\bar{p}}{p}\right)^{\delta}.$$

There is something very counter-intuitive about this result. When 10 out of 10 jurors pass a guilty verdict, we would take this to be strong evidence that the suspect is guilty. When 55 out of 100 jurors pass a guilty verdict, then we would not think much of this narrow proportional margin. The concurring reports in the former case increase our confidence in the reliability of the jurors, whereas the diverging reports in the latter case decrease our confidence in the reliability of the jurors. But this is precisely what is not permitted in List's model, since $\delta = 10$ in both cases. The idea is that 45 guilty votes versus 45 innocent votes brings you precisely back to your prior probability that the suspect is guilty. And, on List's model, your confidence in the reliability of the voters remains unaffected. Hence, the additional 10 guilty votes have precisely the same evidential value as 10 guilty votes when there are only 10 voters.

So how can we model our intuitive reasoning, with its sensitivity to the fact that a 10 versus 0 vote suggests reliable voters, while a 55 versus 45 vote suggests relatively unreliable voters? First, consider the model in the Bayesian Network in Figure 3.5 with reliability defined endogenously and independence between the reliability levels of the jurors. This model is not helpful, since it effectively yields the same results as the model in Figure 3.11 with reliability defined exogenously. To see this, first notice that we have shown in (3.39) and (3.40) that the values $p = \mathbf{P}(\mathrm{REP_i}|\mathrm{HYP})$ and $q = \mathbf{P}(\mathrm{REP_i}|\neg\mathrm{HYP})$ for the model in

Figure 3.5 can be defined in terms of $\rho = \mathrm{P(REL_i)}$ and $a = \boldsymbol{P}(\mathrm{REP_i}|\mathrm{HYP}, \neg\mathrm{REL_i}) = \boldsymbol{P}(\mathrm{REP_i}|\neg\mathrm{HYP}, \neg\mathrm{REL_i})$ for $i = 1, \ldots, n$. It then follows from (3.39) and (3.40) that

$$(3.47) \qquad \rho = p - q \quad \text{and} \quad a = \frac{q}{\bar{p} + q}.$$

From (3.44) and (3.47) it follows that $a = 1/2$ and that $p = (1 + \rho)/2$. We can now rewrite the likelihood ratio x in (3.46) as

$$(3.48) \qquad x = \left(\frac{1 - \rho}{1 + \rho}\right)^{\delta}.$$

Hence, our endogenous reliability model in Figure 3.5 is interdefinable with List's model and will yield the same results.

Second, consider the model with extreme positive relevance between the reliability levels of the jurors, represented in Figure 3.6. With this set-up, the posterior probability that the hypothesis is true strongly exceeds the prior probability, subsequent to a 10 versus 0 vote. But on a 55 versus 45 vote the posterior probability remains equal to the prior. This is easy to understand. If the jurors are either all reliable or all unreliable, then their reports carry evidential weight only as long as they concur. Any divergence is sufficient to conclude that they are *all* unreliable. Since nothing can be learned from jurors who consult a randomizer rather than the real world, the posterior stays at the level of the prior. This is not an unwelcome result in the case at hand. The problem is that the same result holds when there are 99 votes for guilty and 1 vote for innocent. If there is extreme positive relevance between the reliability levels of the jurors, then any dissension is sufficient ground for concluding that the jurors must all be unreliable. This model does not capture our ordinary reasoning about jury votes either.

What we need is a model with moderate positive relevance between the reliability levels of the jurors. It is easy to see that in order to respect (3.44), we need to set the randomization parameter a in the Bayesian Network in Figure 3.7 at $1/2$.[9] We set the values s at .9, t at .1,

[9] This follows directly from (3.39), (3.40), and (3.44).

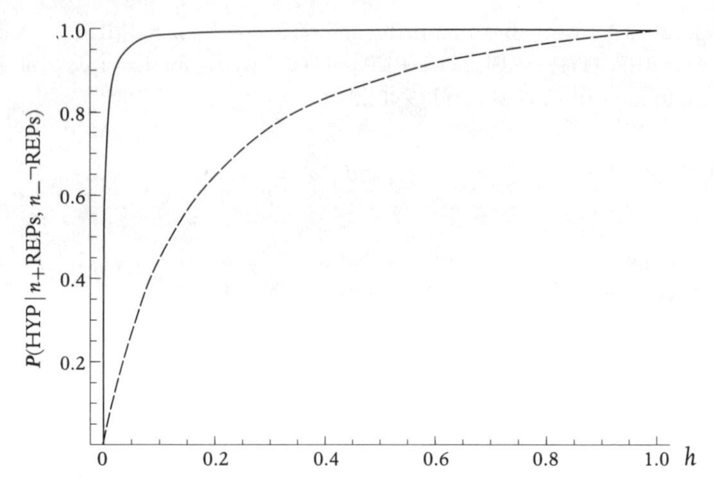

Fig. 3.12 The posterior probability of the hypothesis after 10 positive and 0 negative reports (full line) and after 55 positive and 45 negative reports (dashed line) for moderately positively relevant witnesses with $s = .9$, $t = .1$, and $u = .2$

and u at .2, which brings the prior probability of the reliability of a single juror at $su + t\bar{u} = .2$. In Appendix C.5, we have calculated the posterior probability as a function of the prior probability h that the hypothesis is true for the case in which we receive n_+ guilty votes and n_- innocent votes, for $n_+ > n_-$ (i.e. for $\delta > 0$). We have plotted the case in which we receive 10 guilty votes and 0 innocent votes and the case in which we receive 55 guilty votes and 45 innocent votes in Figure 3.12. In the former case the posterior probability that the information is true steeply rises as a function of the prior probability h. This is precisely what we would expect after concurring votes. But in the latter case the posterior probability just slightly exceeds the prior probability of the hypothesis. Again, this is in line with our intuitions. With so much divergence, we come to distrust the voters, and a balance of 10 votes in favour of a guilty verdict counts for little. Hence, our model of moderate positive relevance captures our ordinary reasoning about such jury votes.[10]

[10] One might object that we often do not know whether the witnesses are genuinely independent. If we receive multiple reports containing the same implausible information, we might want to revise our opinion about the independence of the witnesses. Indeed, in our model, whether there is independence or positive relevance between the reliability

3.7. TVERSKY AND KAHNEMAN'S LINDA

The subjects in Tversky and Kahneman's well-known experiment are told that a certain Linda is 31 years old, single, outspoken, and very bright. She studied philosophy as a student and she was deeply concerned with issues of discrimination and social justice and also participated in anti-nuclear demonstrations. Subsequently they are asked to rank a set of claims about Linda according to what they take to be more likely. The set contains two claims of interest, viz. (i) that Linda is a bank teller and (ii) that Linda is a bank teller (and) active in the feminist movement. A large proportion of the subjects consider the latter claim to be more likely than the former claim. Tversky and Kahneman conclude that the subjects commit the conjunction fallacy, i.e. they believe a conjunction of claims to be more likely than at least one of the conjuncts, which is a clear violation of the Kolmogorov axioms.

But is it not too hasty to conclude that our subjects are irrational? When are people asked to judge whether a proposition is more or less likely to be true? In everyday life, people are typically asked to judge whether a proposition is more or less likely to be true when they have been informed of this proposition by a source (a newspaper, an acquaintance) that may or may not be fully reliable. So it is reasonable to assume that the subjects actually respond to the following question. Would you find it more likely that claim (i) is true given that some partially reliable source informs you that (i), or that claim (ii) is true given that some partially reliable source informs you that (ii)? There is no reason why the latter conditional probability must be lower than the former. We construct a model to determine under what conditions the latter conditional probability is greater than the former.

There are two fact variables, B and F. B can take on the value B, i.e. that Linda is a bank teller, and ¬B. F can take on the value F, i.e. that Linda is active in the feminist movement, and ¬F. There are two report variables, REP_B and REP_F, and a reliability variable, REL, with the standard interpretations. We construct the Bayesian Network in

levels of the witnesses is defined exogenously. One could also take up the challenge of constructing a model in which the independence or the positive relevance between the reliability levels of the witnesses is defined endogenously and changes as a function of the incoming information. This challenge will need to await another occasion.

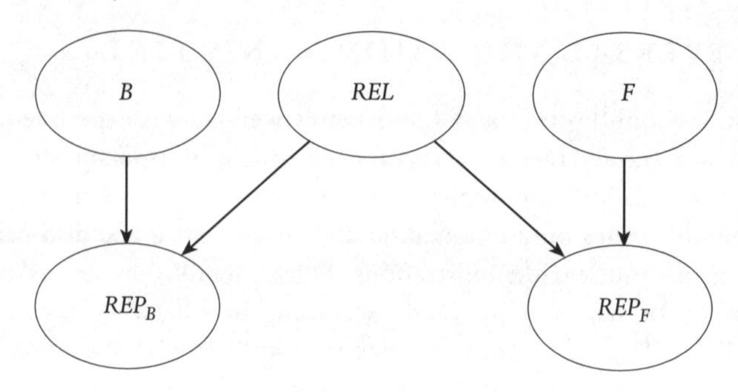

FIG. 3.13 A Bayesian Network for Tversky and Kahneman's Linda

Figure 3.13. Note that, considering the background information that we have received about Linda, $\boldsymbol{P}(\mathrm{B}) = b < \boldsymbol{P}(\mathrm{F}) = f$.

We have introduced the idealization that the variables B and F are independent. If we learn the rather unlikely proposition that Linda is a bank teller, then this does not affect the chance that she is active in the feminist movement, and if we learn the rather unlikely proposition that Linda is active in the feminist movement, then this does not affect the chance that she is a bank teller. This is somewhat unrealistic. One might argue that if Linda has become the kind of person who is willing to work as a bank teller then she is less likely to be active in the feminist movement. Similarly, if Linda has become the kind of person who is active in the feminist movement, then she is less likely to be a bank teller. It is easy to remedy this by introducing negative relevance between B and F, but this just needlessly complicates the model.

The subjects who find claim (ii) more plausible than claim (i) are rational if and only if

$$\Delta P = \boldsymbol{P}(\mathrm{B}, \mathrm{F}|\mathrm{REP_B}, \mathrm{REP_F}) - \boldsymbol{P}(\mathrm{B}|\mathrm{REP_B}) > 0. \tag{3.49}$$

We show in Appendix C.6 that this inequality holds just in case

$$a^2\bar{f} + \bar{a}\rho(a - f(a + \bar{b})) < 0. \tag{3.50}$$

We set both the reliability parameter ρ and the randomization parameter a at .5 and construct a phase graph for the parameters b and f in Figure 3.14. We are only interested in the area underneath the diagonal

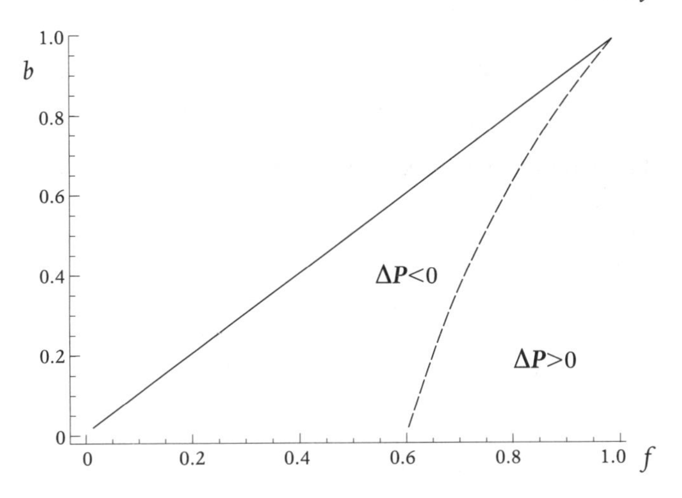

FIG. 3.14 $\Delta P > 0$ if and only if the posterior probability that Linda is a feminist and a bank teller is greater than the posterior probability that Linda is a bank teller for $\rho = .5$ and $a = .5$

in which $f > b$, since given our background information, it is indeed more plausible that Linda is a feminist than a bank teller. Now in the area underneath the phase curve, where f is substantially larger than b, the posterior probability that Linda is a bank teller and a feminist is greater than the posterior probability that she is a bank teller.

How are we to interpret these results? Suppose we are in a courtroom and the judge has provided us with reliable background information about Linda. In case one, a witness is called to the stand, is asked whether Linda is a bank teller, and responds affirmatively. In case two, the same witness is called to the stand, is asked whether Linda is a feminist and whether Linda is a bank teller and responds twice affirmatively. There is a fifty-fifty chance that the witness is reliable and, if unreliable, there is a fifty-fifty chance that she responds affirmatively to each question. Suppose that we think that it is substantially more likely that Linda is a feminist than that she is a bank teller. Then we are prone to think that the affirmative response in case one is due to the witness's unreliability. But in case two, the witness's affirmative response to the question whether Linda is a feminist raises our confidence in her reliability, which raises our confidence in her affirmative response to the question whether she is a bank teller.

The same idea is at work when we ask a partially reliable instrument a number of questions to which we already know the answer before proceeding to the question to which we do not know the answer. Certainly we are more confident that this series of answers is true than had we just asked the question to which we do not know the answer. And it makes no difference whether the instrument freely provides us with the information or just responds to our questions. An instrument that consistently yields information that we already knew to be true and now yields an item of information that is unexpected is to be trusted more than an instrument that just yielded an item of information that is unexpected. In the same vein, we conjecture that even more subjects would find it more probable that (iii) Linda owns a copy of the *Communist Manifesto*, has a poster of Che Guevara in her room, is an admirer of Don Helder Camara, and is a bank teller than that (i) Linda is a bank teller. Of course, this is just armchair cognitive psychology so far. Our analysis needs to be validated by a careful experimental check of these conjectures.

4

Confirmation

4.1. HYPOTHESIS TESTING

So far we have addressed the question of how to deal with information from partially reliable sources in the context of epistemology. The question is also relevant in philosophy of science. What can be learned from experimental results that confirm a hypothesis when we cannot be sure that the instrument is not malfunctioning, that the data analysis techniques are reliable, etc? To answer this question, we slightly extend the basic model of Figure 3.2 in the previous chapter. Hypotheses are typically not tested directly, but rather by examining whether certain consequences would hold if the hypothesis were true. We start with a very simple situation: a hypothesis, a testable consequence of the hypothesis, an instrument that may not be reliable, and a report from this instrument as to whether the testable consequence holds. To model this situation we need four binary propositional variables. The variables *HYP* and *REL* are as defined earlier. The variable *CON* can take on two values: CON—the testable consequence holds—and ¬CON—the testable consequence does not hold. The variable *REP* can take on two values: REP—there is a report that the testable consequence holds—and ¬REP—there is a report that the testable consequence does not hold.[1]

The Bayesian Network for this situation is shown in Figure 4.1. It is the same as the Bayesian Network that modelled the reliability of the witness as an endogenous variable in the previous chapter, with the addition of a node for the variable *CON*. This is our basic model for this chapter. We will take a broad view of what constitutes a testable

[1] Earlier on, ¬REP (and ¬REPR_i) stood for there being no report to the effect that some proposition holds. In this chapter we assume that the instrument provides either a report to the effect that the testable consequence holds or a report to the effect that the testable consequence does not hold. Nothing hinges on the scope of the negation operator for our analysis.

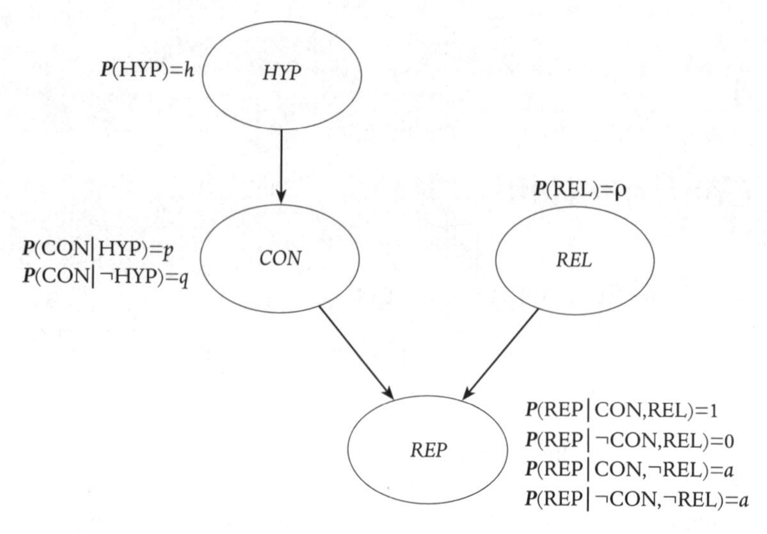

FIG. 4.1 The basic model for testing a hypothesis with a relatively unreliable instrument

consequence. That is, we do not require that the truth of the hypothesis is either a necessary or a sufficient condition for the truth of the testable consequence. Rather, a testable consequence is to be understood as follows. The probability of the consequence given that the hypothesis is true is greater than the probability of the consequence given that the hypothesis is false:

$$P(\text{CON}|\text{HYP}) = p > q = P(\text{CON}|\neg\text{HYP}). \tag{4.1}$$

A testable consequence functions as a partially reliable witness who speaks to the truth of the hypothesis. One may think of a special case in which *CON* states whether a symptom that is a more or less decisive indicator of a disease is present or not and *HYP* states whether the patient does or does not have the disease. The parameter p expresses the true positive rate and the parameter q expresses the false positive rate of the symptom as an indicator for the disease. A medical instrument is brought in to assess whether the symptom is present. *REP* states whether this assessment is positive or negative. *REL* states whether the assessment comes from a reliable or an unreliable instrument. A reliable instrument is completely accurate whereas an unreliable instrument acts like a randomizer with the randomization parameter a.

Let us take a look at the conditional independences that can be read off the model by means of the Parental Markov Condition:

$$HYP \perp\!\!\!\perp REL, \tag{4.2}$$

$$CON \perp\!\!\!\perp REL|HYP, \tag{4.3}$$

$$REP \perp\!\!\!\perp HYP|REL, CON. \tag{4.4}$$

The independence in (4.2) states that if one does not know any values of the variables, then coming to learn that the instrument is reliable or that the instrument is unreliable does not alter the prior probability that the hypothesis is true. This is a plausible assumption as long as one's reasons for believing that the instrument is reliable are independent of the truth of the hypothesis. In Section 4.5, we will investigate what happens when this assumption is violated.

The conditional independence in (4.3) states that if one knows no more than that the hypothesis is true or that the hypothesis is false, then coming to learn in addition that the instrument is reliable or that it is unreliable does not alter the probability that the testable consequence holds. As long as one does not know what report the instrument provides, coming to learn about its reliability teaches us nothing about the testable consequence.

The conditional independence in (4.4) states that if one knows no more than that some definite values of *REL* and *CON* are instantiated, then coming to learn in addition that some definite value of *HYP* is instantiated does not alter the probability of REP. The chance that the instrument will yield a positive or a negative report is fully determined by whether the instrument is reliable and whether the testable consequence holds. Once this information is known, the truth or falsity of the hypothesis itself becomes irrelevant. The latter two independence assumptions seem beyond reproach.[2]

[2] The Bayesian Network also represents a series of other conditional independences, e.g. $REP \perp\!\!\!\perp HYP|CON$. These independences can be derived by means of the semi-graphoid axioms or can be read off by the d-separation criterion. Compare fn. 6 in Chapter 3.

We are interested in the probability of the hypothesis given that there is a report that the testable consequence holds. Following the procedure outlined in the previous chapter, we show in Appendix D.1 that

$$(4.5) \quad \mathbf{P}^*(\mathrm{HYP}) = \mathbf{P}(\mathrm{HYP}|\mathrm{REP}) = \frac{h}{h + \overline{h}x_0} \text{ with the likelihood ratio } x_0 = \frac{q\rho + a\overline{\rho}}{p\rho + a\overline{\rho}}$$

with the parameters h, p, q, and ρ as specified in the Bayesian Network in Figure 4.1. Note that $0 < x_0 < 1$ for $p > q$. We plot $\mathbf{P}^*(\mathrm{HYP})$ as a function of the reliability of the instrument ρ in Figure 4.2. This function is a monotonically increasing function. Furthermore, we show in Appendix D.2 that:

$$(4.6) \quad \mathbf{P}^*(\mathrm{HYP}) \text{ is } \begin{array}{l} \text{concave for } a < c, \\ \text{linear for } a = c, \text{ and} \\ \text{convex for } a > c \end{array} \quad \text{for } c = \mathbf{P}(\mathrm{CON}) = hp + \overline{h}q.$$

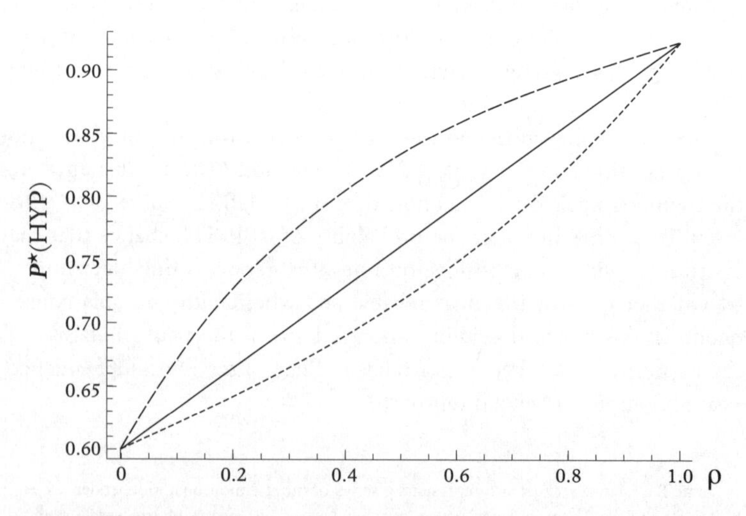

FIG. 4.2 The posterior probability that the hypothesis is true as a function of the reliability parameter ρ when the randomization parameter is set at $a = .52$ (full line), $a = .2$ (dotted line) and $a = .9$ (dashed line) and the prior probability of the testable consequence is $c = .52$

Intuitively, one's degree of confidence in a hypothesis conditional on the report rises with a decreasing value of the randomization parameter a. For small a, the chance of a positive report coming from an unreliable instrument is small. For large a, the chance of a positive report coming from an unreliable instrument is large. Only positive reports from reliable instruments increase our degree of confidence in the hypothesis. Hence, the curves for lower values of the randomization parameter are above the curves for higher values of the randomization parameter.

There are various measures of confirmation in the literature (for a survey, see Fitelson (1999 and 2001) and Eells and Fitelson (2001)). We investigate the difference measure, i.e. the difference between the posterior and the prior probability of the hypothesis:[3]

$$P^*(\text{HYP}) - P(\text{HYP}) = \frac{h\bar{h}\bar{x}_0}{h + \bar{h}x_0} \tag{4.7}$$

with the likelihood ratio x_0 as defined in (4.5).

We have shown how to model the degree of confirmation that a hypothesis receives from a single positive report concerning a single testable consequence of the hypothesis. There are various routes one can take to improve one's degree of confidence when working with instruments whose reliability is in doubt. First, one can perform multiple measurements of the same testable consequence to check whether the same test results obtain. One can do so with the same test instrument or with different test instruments. Second, one can seek to test multiple testable consequences of the hypothesis. Again, this can be done with the same test instrument or with different test instruments. Third, one can seek support for the reliability of the instrument by appealing to an *auxiliary theory* that may or may not be dependent on the hypothesis under investigation.

Our investigation of the first two routes will have some surprising consequences for the variety-of-evidence thesis, i.e. the thesis that more varied evidence in support of a hypothesis provides more confirmation than less varied evidence, ceteris paribus. We interpret more varied evidence as evidence that stems from multiple instruments

[3] For a discussion of how other measures of confirmation fare in the present context, see Bovens and Hartmann (2002: 70–1).

(rather than a single instrument) and that tests multiple testable consequences (rather than a single testable consequence) of the hypothesis. Our investigation of the third route is relevant to the Duhem–Quine thesis, i.e. the thesis that hypotheses are not tested in isolation but always on the background of auxiliary theories. In particular, we will provide a Bayesian analysis of confirmation from test results in cases where the hypothesis and the auxiliary theory are *not* independent.

4.2. SAME TEST RESULTS

Suppose that we have tested a single testable consequence of a hypothesis by means of a single instrument. We have received a positive report, but we want to have additional confirmation for our hypothesis. There are two possibilities. Either we can take our old instrument and run the test a couple more times. Or we can choose new and independent instruments and test for the very same testable consequence with these new instruments. First, we will show that both of these routes can be successful. If we receive more reports to the effect that the testable consequence holds, either from our old instrument or from new and independent instruments, then the hypothesis does indeed receive additional confirmation. Second, we are curious to know which is the better route, assuming that we do indeed receive more reports to the effect that the testable consequence holds. In other words, which route yields a higher degree of confirmation? Is there a univocal answer to this question, or is one route more successful under certain conditions, while the other route is more successful under other conditions? To keep things simple, we will present our analysis for *one* additional test report, either from the same or from different instruments.

Let us first model the degree of confirmation that the hypothesis receives from an additional positive report from the same instrument. Figure 4.3 represents this situation. Starting with our basic model in Figure 4.1, Figure 4.3 results from substituting REP_1 for REP and adding a node to represent the binary variable REP_2. Just like REP_1, REP_2 is directly influenced by REL and CON, and so two more arrows are drawn. We impose a symmetry condition on the probability measure $\mathbf{P}$ for this graph, i.e. we assume for the second report that the instrument is either fully reliable or fully unreliable with the same randomization parameter a.

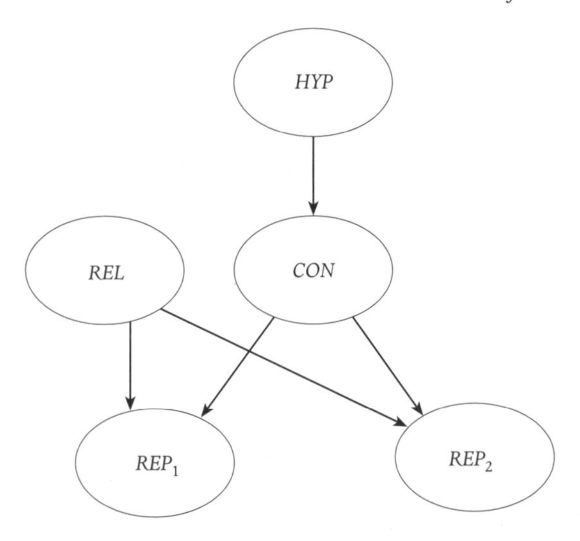

FIG. 4.3 Multiple measurements with a single instrument of a single testable consequence

Second, we model the degree of confirmation that the hypothesis receives from an additional confirming report from a second independent instrument. Figure 4.4 results from adding nodes for the new instrument to our basic model in Figure 4.1. We substitute REL_1 for *REL* and REP_1 for *REP* and add a node for the variable REL_2, which expresses whether the second instrument is reliable or not. We add a node for the variable REP_2, which expresses whether the second instrument provides a report that the testable consequence holds or a report that it does not hold. REP_2 is directly influenced by REL_2 and *CON*, so we draw in two more arrows. To keep matters simple, we impose a condition of symmetry on the probability measure $\mathbf{P}'$ for this graph: There is an equal chance ρ that both instruments are reliable, and if the instruments are unreliable then they randomize at the same level a.[4] To compare the situation with one instrument to the situation with two instruments we need to keep other factors the same in both cases.

[4] Our model does not apply to unreliable instruments that do not randomize, but rather provide accurate measurements of other features than the features they are supposed to measure. In effect, our model exploits the coherence of the reports as an indicator that the reports are obtained from reliable rather than unreliable instruments. But if unreliable

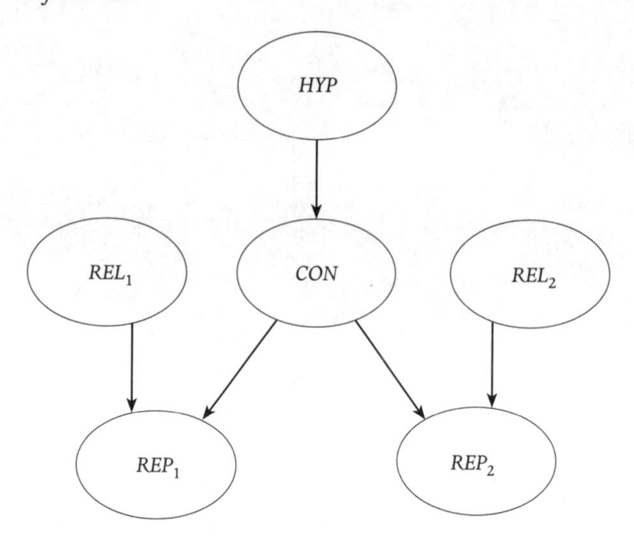

FIG. 4.4 Measurements with multiple instruments of a single testable consequence

For this reason we impose the same values of the parameters a, h, p, q, and ρ on the probability measures $\mathbf{P}$ and $\mathbf{P}'$.

The two instruments are independent of one another. What this means is that

$$REP_i \perp\!\!\!\perp REP_j | CON \; \forall i, j = 1, 2 \text{ and } i \neq j. \tag{4.8}$$

Given that the testable consequence holds (or that it does not hold), there is a certain chance that we will receive a report to the effect that the consequence holds. Whether or not we receive another report to this effect does not affect this chance. An independent instrument may not always provide us with an accurate report, but it is not influenced by what other instruments report. The conditional independence (4.8) can be read off from the graph in Figure 4.4.

We show in Appendix D.3 that obtaining a concurring report from the same instrument about the same testable consequence always provides additional confirmation to the hypothesis:

instruments accurately measure features other than the ones they are supposed to measure, then they will also provide coherent reports and so the coherence of the report is no longer an indicator that they were obtained from reliable instruments.

(4.9) $$\Delta \boldsymbol{P} = \boldsymbol{P}(\mathrm{HYP}|\mathrm{REP}_1, \mathrm{REP}_2) - \boldsymbol{P}(\mathrm{HYP}|\mathrm{REP}_1) > 0.$$

We also show in Appendix D.4 that obtaining a concurring report from a different instrument about the same testable consequence always provides additional confirmation to the hypothesis:

(4.10) $$\Delta \boldsymbol{P} = \boldsymbol{P}'(\mathrm{HYP}|\mathrm{REP}_1, \mathrm{REP}_2) - \boldsymbol{P}'(\mathrm{HYP}|\mathrm{REP}_1) > 0.$$

We now turn to the question whether the hypothesis receives more confirmation from a concurring report from one and the same instrument or from independent instruments. We show in Appendix D.5 that

(4.11) $$\Delta \boldsymbol{P} = \boldsymbol{P}'(\mathrm{HYP}|\mathrm{REP}_1, \mathrm{REP}_2) - \boldsymbol{P}(\mathrm{HYP}|\mathrm{REP}_1, \mathrm{REP}_2) > 0 \text{ iff } 1 - 2\bar{a}\,\bar{\rho} > 0.$$

The graph in Figure 4.5 represents this inequality. For values of (ρ, a) above the phase curve, $\Delta \boldsymbol{P} > 0$—i.e. it is better to receive reports from two instruments. For values of (ρ, a) on the phase curve, $\Delta \boldsymbol{P} = 0$—i.e. it makes no difference whether we receive reports from one or two instruments. For values of (ρ, a) below the phase curve, $\Delta \boldsymbol{P} < 0$—i.e. it is better to receive reports from one instrument than from two instruments.

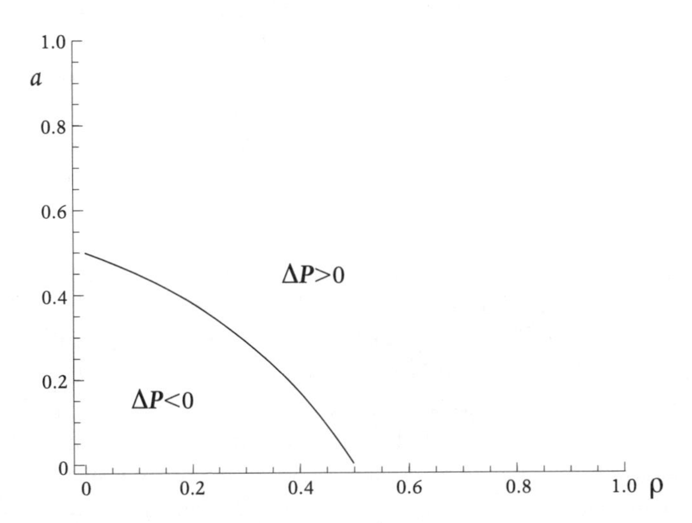

FIG. 4.5 $\Delta P > 0$ if and only if positive reports from two instruments testing the same testable consequence yield more confirmation to the hypothesis than from a single instrument

Do these results seem plausible? There are two conflicting intuitions at work here. On the one hand, we are tempted to say that concurring reports from two instruments is the better way to go, since independence is a good thing. On the other hand, if we receive concurring reports from a single instrument, then we feel more confident that the instrument is not a randomizer and this increase in confidence in the reliability of the instrument benefits the confirmation of the hypothesis. For higher values of ρ, the former consideration becomes more weighty than the latter. There is not much gain to be made in our confidence in the reliability of the instrument by using it over again and so we might as well enjoy the benefits of two independent tests. For lower values of a, the latter consideration becomes more weighty. If we are working with an instrument which, if unreliable, has a low chance of providing positive reports, then concurring reports constitute a substantial gain in our confidence in its reliability, which in turn benefits the confirmation of the hypothesis.

4.3. COHERENT TEST RESULTS

The second route to raise the degree of confirmation for a hypothesis is to identify a *range* of testable consequences that can be assessed by either a single or by multiple independent instruments. Let us draw the graphs for two testable consequences. Following our heuristic, the hypothesis (*HYP*) directly influences the testable consequences (CON_i for $i = 1, 2$). Figure 4.6 represents the case in which there is a single instrument. Each testable consequence (CON_i) conjoined with the reliability of the single instrument (*REL*) directly influences the report about the testable consequence in question (REP_i). Figure 4.7 represents the case in which there are two independent instruments. Each testable consequence (CON_i) conjoined with the reliability of the instrument that tests this consequence (REL_i) directly influences the report about the testable consequence in question (REP_i). We define a probability measure $\boldsymbol{P}$ for the Bayesian Network in Figure 4.6 and a probability measure $\boldsymbol{P'}$ for the Bayesian Network in Figure 4.7. We impose the symmetry condition on each measure and impose the same values of the parameters a, h, p, q, and ρ on the probability measures $\boldsymbol{P}$ and $\boldsymbol{P'}$ to permit a comparison between both situations.

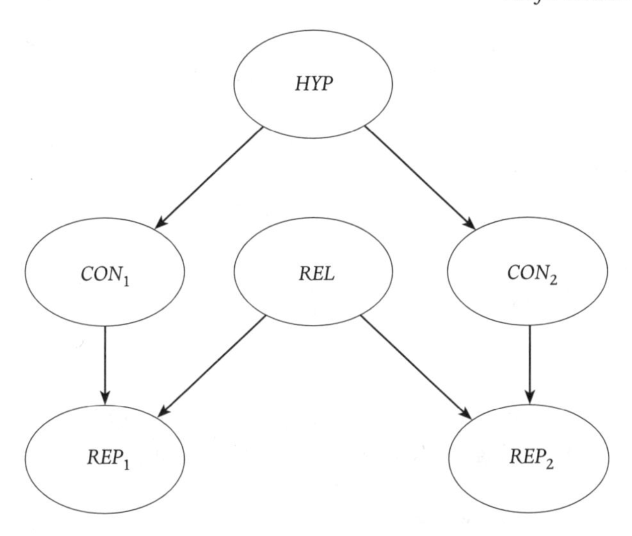

FIG. 4.6 Measurements with a single instrument of multiple testable consequences

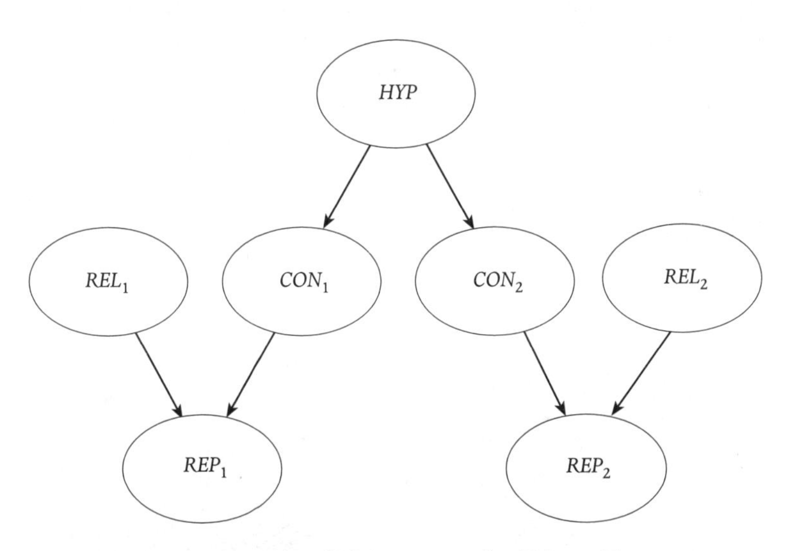

FIG. 4.7 Measurements with multiple instruments of multiple testable consequences

We show in Appendix D.6 that a concurring report always raises the degree of confirmation with multiple instruments:

$$\Delta \boldsymbol{P} = \boldsymbol{P}'(\text{HYP}|\text{REP}_1, \text{REP}_2) - \boldsymbol{P}'(\text{HYP}|\text{REP}_i) > 0, \tag{4.12}$$

whereas with a single instrument, this is not the case, as we show in Appendix D.7:

$$\Delta \boldsymbol{P} = \boldsymbol{P}(\text{HYP}|\text{REP}_1, \text{REP}_2) - \boldsymbol{P}(\text{HYP}|\text{REP}_1) > 0 \quad \text{iff } pq\rho + a\overline{\rho}(p + q - a) > 0. \tag{4.13}$$

We illustrate this in Figure 4.8, where we fix $a = .5$ and construct phase curves for high-, medium- and low-range values of the reliability parameter ρ. In Figure 4.9, we fix $\rho = .5$ and construct phase curves for high-, medium- and low-range values of the randomization parameter a. Since we have stipulated that $p > q$, we are only interested in the areas below the straight line that represents $p = q$ in both figures.

The areas in these graphs in which $\Delta \boldsymbol{P} < 0$ are rather curious. For certain values of a, p, q, and ρ, we test a first consequence of a hypothesis, receive a positive report, and are more confident that the hypothesis is true. Then we test a second consequence of the hypothesis

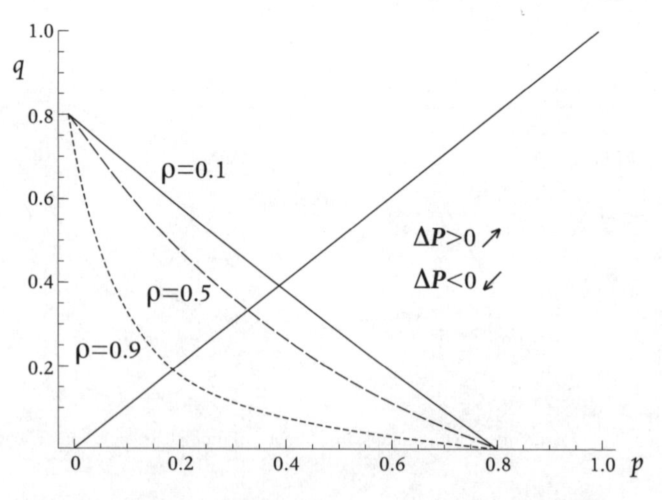

FIG. 4.8 $\Delta P > 0$ if and only if two positive reports from a single instrument testing two testable consequences yield more confirmation to the hypothesis than one positive report testing a single testable consequence for $a = .5$; the relevant region is the region where $p > q$

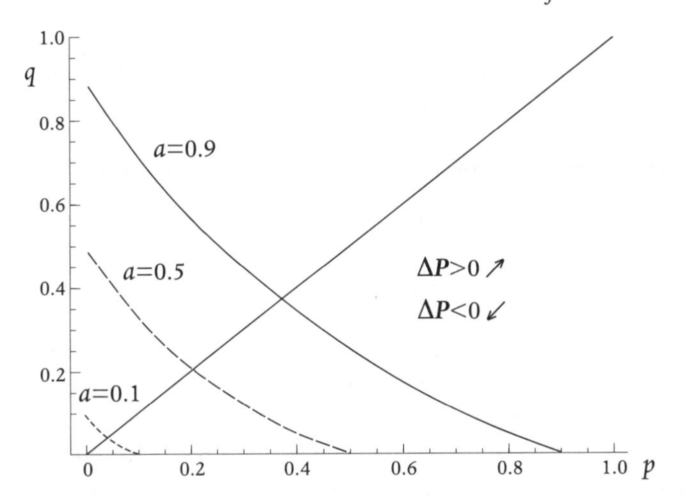

FIG. 4.9 $\Delta P > 0$ if and only if two positive reports from a single instrument testing two testable consequences yield more confirmation to the hypothesis than one positive report testing a single testable consequence for $\rho = .5$; the relevant region is the region where $p > q$

with the very same instrument, receive once again a positive report, but this time round our degree of confidence in the hypothesis drops. How can we interpret these results?

Notice that the effect is most widespread for (i) lower values of ρ, (ii) higher values of a, and (iii) lower values of p. To get a feeling for the magic of the numbers, let us look at a range of values where the effect occurs par excellence. Hence, let us consider instruments that are not likely to be reliable, and, if unreliable, have a high chance of providing a positive report. And let us consider testable consequences that are unlikely to occur when the hypothesis is true (though of course the testable consequences are still more likely to occur when the hypothesis is true than when the hypothesis is false). Considering (i), we do not have much trust in the instrument to begin with. Now it gives us nothing but positive reports. Considering (ii), the instrument is likely to be a randomizer and so we become even more confident that the instrument is unreliable. But should this not be offset by the fact that we receive reports about coherent test results in support of our hypothesis? No, since considering (iii), our tests are rather weak and these coherence effects count for little. Hence, when we get a concurring report, we become very confident that the instrument is unreliable, and consequently our confidence in the hypothesis drops.

We now turn to the question of whether the hypothesis receives more confirmation from a concurring report from one and the same instrument or from independent instruments. We show in Appendix D.8 that

$$(4.14) \qquad \Delta P = P'(\mathrm{HYP}|\mathrm{REP}_1, \mathrm{REP}_2) - P(\mathrm{HYP}|\mathrm{REP}_1, \mathrm{REP}_2) > 0$$
$$\text{iff } (2a - p - q)a - 2(a - p)(a - q)\rho > 0.$$

To evaluate this expression let us assume that the testable consequences are strongly tied to the hypothesis. We do this by fixing $p = .9$ and $q = .1$. We construct a phase curve for values of (a, ρ) in Figure 4.10. If the randomization parameter and the reliability parameter are set low, then one instrument tends to do better than two. Subsequently we assume mid-range values for the randomization and the reliability parameters ($a = .5$ and $\rho = .5$) and construct a phase curve for values of (p, q) in Figure 4.11. We are interested in the area where $p > q$ or, in other words, the area below the straight line. If q is set high, i.e. if the testable consequences occur frequently even when the hypothesis is false, then one instrument tends to do better than two.[5]

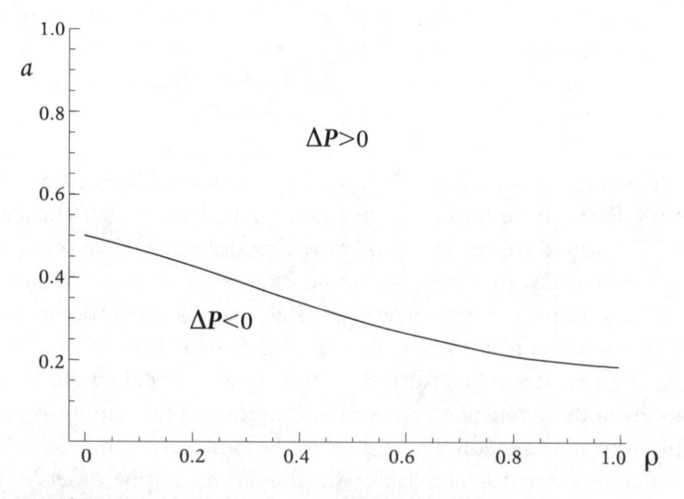

Fig. 4.10 $\Delta P > 0$ if and only if positive reports from two instruments testing two testable consequences yield more confirmation to the hypothesis than positive reports from a single instrument testing two testable consequences for $p = .9$ and $q = .1$

[5] Note that by introducing scaled parameters, $p' = p/a$ and $q' = q/a$, the parameter a can be eliminated from the phase curve conditions in (4.13) and (4.14).

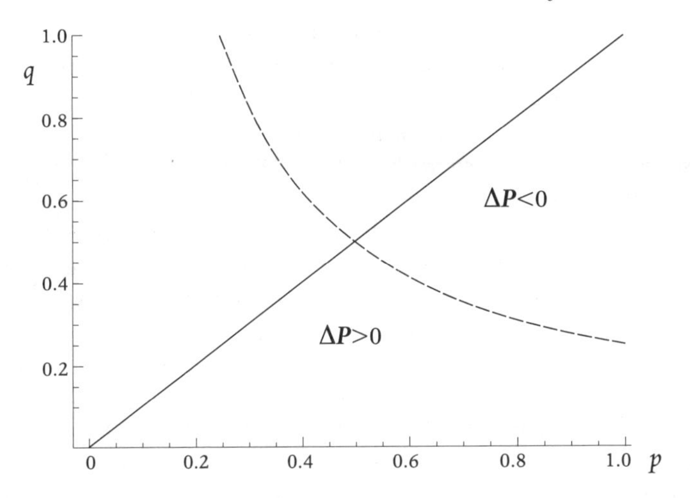

Fig. 4.11 $\Delta P > 0$ if and only if positive reports from two instruments testing two testable consequences yield more confirmation to the hypothesis than positive reports from a single instrument testing two testable consequences for $a = .5$ and $\rho = .5$; the relevant region is the region where $p > q$

In the previous section, we explained why the effect of concurrent reports on the reliability of a single instrument carries more weight than the independence of multiple instruments for lower values of a and ρ. The same explanation holds here. But why is this effect amplified for higher values of q? The higher the values of q, the more likely the testable consequences will occur and so concurring reports will boost our confidence in the reliability of the single instrument even more. Hence higher values of q tend to favor a single instrument over multiple instruments.

4.4. THE VARIETY-OF-EVIDENCE THESIS

An often stated scientific intuition is that more varied evidence will better confirm a hypothesis than less varied evidence, ceteris paribus. The Bayesian explication of this intuition is a textbook case of a Bayesian success story. It is shown that the increment of confirmation received by a hypothesis from confirming test results becomes smaller and smaller as the same old test is run over and over again (e.g.

Earman (1992: 77–9) and Howson and Urbach (1993: 119–23)). But what does it mean for the same old test to be run over and over again? We could take it to mean that one repeatedly checks the same old testable consequence rather than that one checks multiple independent testable consequences of the hypothesis. Or we could take it to mean that one repeatedly checks a testable consequence with the same old instrument rather than with multiple independent instruments.

How should we understand the ceteris paribus clause in the variety-of-evidence thesis? Let there be two sets of evidence, one containing a less varied pair and one containing a more varied pair of items of evidence. The ceteris paribus clause is satisfied if each item of evidence i within these sets j has the same evidential strength, as expressed by the likelihood ratio $\boldsymbol{P}(\mathrm{REP}_i^j|\neg\mathrm{HYP})/\boldsymbol{P}(\mathrm{REP}_i^j|\mathrm{HYP})$.[6] The following analogy supports this interpretation. Suppose one wants to test the claim that a varied set of investments promises a greater yield than a non-varied set. Then it would be meaningless to compare a varied set of investments that each have a high rating with a non-varied set of investments that each have a low rating or vice versa. A meaningful comparison can be made if the individual investments in the varied set of investments have the same ratings as the individual investments in the non-varied set of investments. Keeping the evidential strength of each item of evidence fixed within the respective sets of tests is like using investments with the same ratings. It is easy to prove that this ceteris paribus clause is satisfied in our comparisons since the values of the parameters a, h, p, q, and ρ are held fixed.

Our analysis permits us to impose the following *caveats* on the variety-of-evidence thesis. We showed in Section 4.2 that,

(i) if we are testing a single testable consequence, it is sometimes more beneficial for the confirmation of the hypothesis to receive positive reports from the same instrument than from different instruments, ceteris paribus.

In section 4.3 we have shown that,

(ii) if we are testing different testable consequences, it is sometimes more beneficial for the confirmation of the hypothesis to receive positive reports from the same instrument than from different instruments, ceteris paribus.

[6] This is a sufficient condition. It is more difficult to spell out a necessary condition, since one could argue that a comparison between two sets of evidence is also meaningful when both sets contain items of evidence with a similar mixture of evidential strengths.

And there is still another conclusion to be drawn from our results. We saw in Section 4.2 that when instruments may not be reliable it is always a good thing for the confirmation of the hypothesis to receive a concurring report from the same instrument about the same testable consequence. But in Section 4.3 we saw that our confidence in the hypothesis may decrease as we receive a concurring report from the same instrument about a different testable consequence. Hence, we can add a third caveat:

(iii) If we are testing with a single instrument, it is sometimes more beneficial for the confirmation of the hypothesis to receive positive reports about the same testable consequence than about different testable consequences, ceteris paribus.

There are two Bayesian approaches to the problem of the variety of evidence present in the literature. On the *correlation approach* the items of evidence $E_1, \ldots, E_n$ are considered less varied when there is a greater rate of increase in the probability values $\boldsymbol{P}(E_1)$, $\boldsymbol{P}(E_2|E_1)$, $\ldots$, $\boldsymbol{P}(E_n|E_1, \ldots, E_{n-1})$ (Howson and Urbach (1993: 119–23); Earman (1992: 77–9)). On the *eliminative approach* evidence E in support of the hypothesis H_i is considered more varied when the likelihoods of E on hypotheses H_j that are alternatives to H_i—i.e. $\boldsymbol{P}(E|H_j)$ (for $j = 1, \ldots, i-1, i+1, \ldots, n$)—have lower values. In other words, more varied evidence is evidence that permits us to exclude more competing hypotheses (Horwich (1982: 118–22) and Wayne (1995: 116)). Each of these approaches starts from a particular pre-theoretical intuition about variety. Our approach does no less. The pre-theoretical intuition that we start with is that evidence that proceeds from multiple instruments and that addresses multiple testable consequences is more varied than evidence that proceeds from a single instrument or that addresses a single testable consequence.

How does our analysis compare to the correlation approach? Since we keep the values of the parameters a, h, p, q, and ρ fixed, it follows that $\boldsymbol{P}(REP_1) = \boldsymbol{P}'(REP_1)$ in all of our comparative cases. Hence, on the correlation approach, the evidence is less varied the more the conditional probability of REP_2 given REP_1 exceeds the marginal probability of REP_1. It is easy to show that the former exceeds the latter to a greater extent when testing with a single instrument rather than with multiple instruments and when testing a single testable consequence rather than multiple testable consequences. However, our analysis shows that evidence being less varied as defined on the correlation approach is no guarantee that the hypothesis will receive less confirmation. For instance, consider the cases that are modelled by the graphs in Figures 4.3 and 4.4. Set $a = .2$, $h = .5$, $p = .9$, $q = .1$, and $\rho = .2$ in both measures. Then $P(\text{REP}_1) = P'(\text{REP}_1) = .26$, but there is a stricter correlation and hence less variety of evidence when the reports come from a single instrument than from two independent instruments, viz. $P(\text{REP}_2|\text{REP}_1) \approx .51 > .30 \approx P'(\text{REP}_2|\text{REP}_1)$. However, the hypothesis receives more confirmation when the reports come from a single rather than from multiple instruments, viz. $P(\text{HYP}|\text{REP}_1, \text{REP}_2) \approx .80 > .77 \approx P'(\text{HYP}|\text{REP}_1, \text{REP}_2)$.

How do our results square with the formal results that allegedly prove the variety-of-evidence thesis on the correlation approach? These formal results rest on the assumption that the evidence is strictly entailed by the hypothesis, viz. $P(\text{E}|\text{H}) = P'(\text{E}|\text{H}) = 1$. This is a restrictive constraint on the notion of evidence, and quite unrealistic in many contexts, such as in the context of the diagnosis of disease. What our examples show is that less varied evidence may indeed provide more confirmation to the hypothesis if we work with a looser notion of evidence and relax the assumption to $P(\text{E}|\text{H}) = p > q = P(\text{E}|\neg\text{H})$.

Let us turn to the eliminative approach. What is the import of the eliminative approach when there are only two hypotheses, viz. H and ¬H, as is the case in our examples? Suppose that we want to ascertain whether a patient in a hospital has Lyme disease (H). One set of evidence E contains vomiting, fever, etc. Another set of evidence E′ contains a recent tick bite, a characteristic rash, etc. It is plausible to set $P(\text{E}|\text{H}) = P(\text{E}'|\text{H})$ and $P(\text{E}|\neg\text{H}) > P(\text{E}'|\neg\text{H})$. Then on the eliminative approach, E′ is more varied than E. Fitelson (1996: 654–6) argues that the eliminative approach requires the additional ceteris paribus assumption that the likelihoods of both *sets* of evidence on the hypothesis are identical, which translates in our case to $P(\text{REP}_1, \text{REP}_2|\text{HYP}) = P'(\text{REP}_1, \text{REP}_2|\text{HYP})$. This ceteris paribus assumption is not satisfied in any of our comparisons. We do not find this disconcerting, since we think that the eliminative notion of variety of evidence is really a stretch of the ordinary notion of variety of evidence. Certainly E′ has more diagnostic value than E. But is this due to it being more diverse? Note that, on the eliminative approach, a single item of evidence could be more varied than some other single item of evidence, which seems somewhat odd. What the eliminative approach seems to capture is how 'diversifying' the evidence is—i.e. what its capability to distinguish between a wide range of competing hypotheses is. Furthermore, even if a case can be made that this notion corresponds to an intuitively plausible notion of variety of evidence, the notion we are trying to capture is rather different from the notion sought by the eliminative approach.

4.5. AUXILIARY THEORIES

Let us return to our basic model of Figure 4.1. In this model, the variable *REL* is a root node and we have assigned a probability value ρ that expresses the chance that the instrument is reliable. It is a common theme in contemporary philosophy of science that the workings of the instrument are themselves supported by an auxiliary theory of the instrument. If this is the case, then a more complete model would not represent *REL* as a root node. Whether the instrument is reliable or not is directly influenced by whether the auxiliary theory holds or not (*AUX*). Just as we assigned a prior probability to the hypothesis, we also assign a prior probability t to the auxiliary theory. To keep matters simple, let us assume in this section that the instrument is reliable just in case the auxiliary theory is correct, and that the testable consequence holds just in case the hypothesis is true. Our basic model is then expanded in the Bayesian Network in Figure 4.12. In this Bayesian Network, *AUX* and *HYP* are still independent. This may or may not be a realistic assumption. Sometimes the auxiliary theory has no relation whatsoever to the hypothesis under test. But sometimes they are quite closely tied to each other. For instance, they may both be parts of a broader theory. We can model positive relevance between *AUX* and *HYP* by connecting both variables in the Bayesian Network and by setting $P'(\text{AUX}|\text{HYP}) = t_h > t_{\bar{h}} = P'(\text{AUX}|\neg\text{HYP})$ as in Figure 4.13.

We can now ask the following question. Ceteris paribus, does the hypothesis receive more or less confirmation if the auxiliary theory that supports the reliability of the instrument is independent rather than positively relevant to the hypothesis under test? To respect the ceteris paribus clause we must make sure that the randomization parameter,

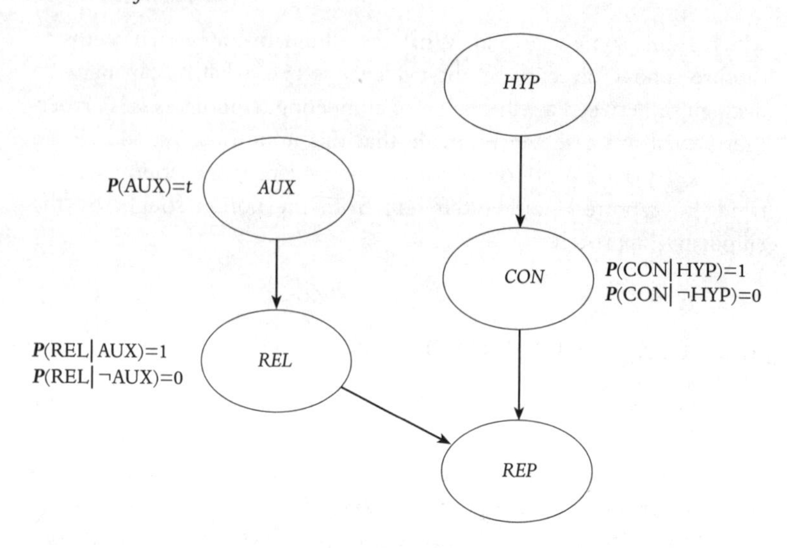

Fig. 4.12 The reliability of the instrument is supported by an independent auxiliary theory

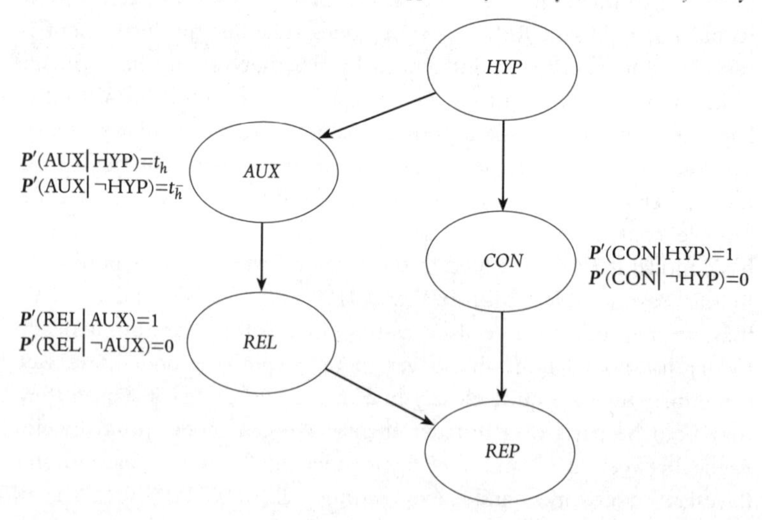

Fig. 4.13 The reliability of the instrument is supported by a positively relevant auxiliary theory

the reliability parameter, and the prior probability of the hypothesis are fixed across both situations. To fix the reliability parameter, we must make sure that $t = t_h h + t_{\bar{h}} \bar{h}$, since the instrument is reliable just in case the auxiliary theory is true. We show in Appendix D.9 that

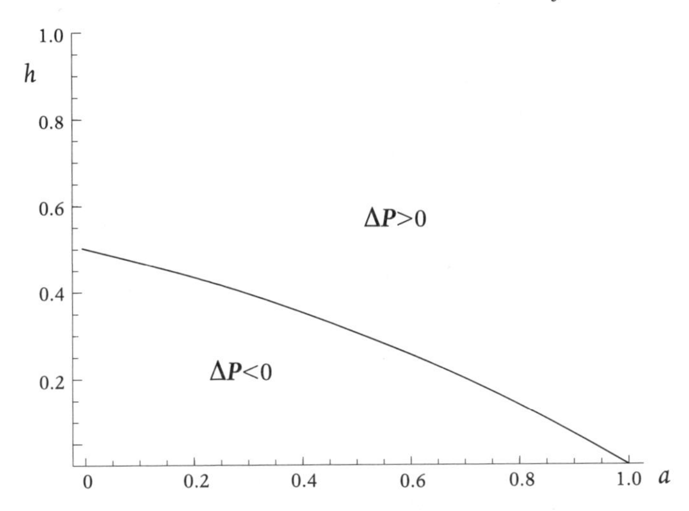

FIG. 4.14 $\Delta P > 0$ if and only if the hypothesis receives additional confirmation when the reliability of the instrument is supported by an independent rather than a positively relevant auxiliary theory with $t_h = .8$ and $t_{\bar{h}} = .2$

$$\text{(4.15)} \qquad \Delta \boldsymbol{P} = \boldsymbol{P}(\text{HYP}|\text{REP}) - \boldsymbol{P}'(\text{HYP}|\text{REP}) > 0$$
$$\text{iff } h + \bar{a}(ht_h + \bar{h}t_{\bar{h}} - 1) > 0.$$

To evaluate this expression, we construct two graphs. In Figure 4.14, we set $t_h = .8$ and $t_{\bar{h}} = .2$ and construct a phase curve for (a, h). In Figure 4.15, we set $a = 1/3$ and $h = 1/3$ and construct a phase curve for $(t_h, t_{\bar{h}})$.

What we see in Figure 4.14 is that a positively relevant auxiliary theory provides more of a boost to the degree of confirmation of the hypothesis from a positive report than an independent auxiliary theory when the prior probability of the hypothesis is relatively low and the value of the randomization parameter is relatively low—and vice versa for relatively high values of these parameters. In Figure 4.15 we are only interested in the area below the line, where $t_h > t_{\bar{h}}$. What we see is that for $t_h < .5$, a positively relevant auxiliary theory always provides more of a boost to the degree of confirmation of the hypothesis. But for $t_h > .5$, a positively relevant auxiliary theory provides more of a boost if and only if the values of $t_{\bar{h}}$ are sufficiently smaller than

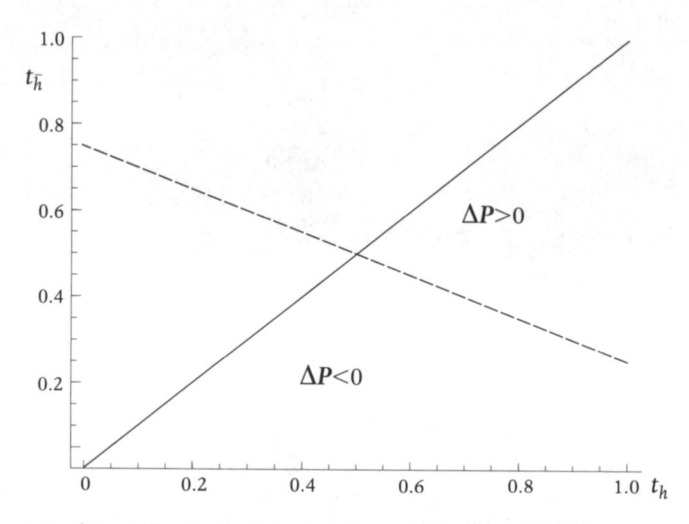

Fig. 4.15 $\Delta P > 0$ if and only if the hypothesis receives more confirmation when the reliability of the instrument is supported by an independent rather than a positively relevant auxiliary theory with $a = 1/3$ and $h = 1/3$; the relevant region is the region where $t_h > t_{\bar{h}}$

t_h—in other words, if and only if the auxiliary theory is sufficiently positively relevant to the hypothesis.

Can an intuitive account be given of these results? Why does a positively relevant auxiliary theory yield a higher degree of confirmation for an implausible hypothesis than an independent auxiliary theory, as indicated by the graph in Figure 4.14? If h is low, say $h = .1$, then a positively relevant auxiliary theory has a low prior probability $t = .8 \times .1 + .2 \times .9 = .26$. The ceteris paribus clause requires that we set the prior probability of an independent auxiliary theory at .26 as well. Since the hypothesis is improbable, it is likely that a positive report is due to the unreliability of the instrument and hence the falsity of the auxiliary theory. For example, for $a = .5$, the posterior probability of the auxiliary theory slides below .26. However, the blame falls much more heavily on the independent auxiliary theory than on the positively relevant auxiliary theory, because the probability of the latter is tied to the probability of the hypothesis. Actually, the posterior probability of the independent auxiliary theory goes into free fall to a value slightly below .07, whereas the posterior probability of the positively relevant auxiliary theory remains at a respectable .17. With this distrust in the auxiliary theory and hence in the instrument it is understandable that the hypothesis will receive less confirmation from a positive report when the instrument is supported by an independent auxiliary theory than when it is supported by a positively relevant auxiliary theory. Actually, the posterior probability of the hypothesis is .20 with a positively relevant auxiliary theory as opposed to a value slightly below .16 with an independent auxiliary theory. This argument will hold whenever the auxiliary theory is sufficiently positively relevant to the hypothesis, as Figure 4.15 illustrates.

The Duhem–Quine thesis notoriously states that if our experimental results are not in accordance with the hypothesis under investigation, there is no compelling reason to reject the hypothesis, since the blame could just as well fall on the auxiliary theories. One virtue of our model is that it gives a precise Bayesian account of how experimental results affect our confidence in the hypothesis and our confidence in the auxiliary theory. But there is also a more important lesson to be learned. In discussing the Duhem–Quine thesis, Bayesians typically assume that the auxiliary theory and the hypothesis are independent (Howson and Urbach (1993: 139)), although there is some cursory discussion of dependence between the hypothesis and the auxiliary theory in Dorling (1996). The assumption of independence certainly makes the calculations more manageable, but it does not square with the holism that is the inspiration for the Duhem–Quine thesis. Not only are experimental results determined by a hypothesis and auxiliary theories, they are determined by a hypothesis and auxiliary theories that are often hopelessly interconnected with each other. And these interconnections raise havoc in assessing the value of experimental results in testing hypotheses. There is always the fear that the hypothesis and the auxiliary theory really come out of the same deceitful family and that the lies of one reinforce the lies of the others. What our results show is that this fear is not entirely ungrounded. For hypotheses with a high prior probability, it is definitely better that the reliability of the instrument be supported by an independent auxiliary theory. But on the other hand, for hypotheses with a low prior probability we should cast off such fears. Hypotheses and auxiliary theories from the same family are very welcome, since positive test reports will provide stronger confirmation to the hypothesis under consideration.

5

Testimony

5.1. THE VALUE OF SURPRISING INFORMATION

We question some independent and partially reliable witnesses about the fate of a person missing in action. Compare the following two cases. First, the witnesses all tell us that the person in question was shot. Second, the witnesses all tell us that the person in question was forced to drink a cup of hemlock. The latter story may strike us as less convincing, since there is nothing out of the ordinary about a person missing in action being shot. It may well be the case that the witnesses independently made up the story without knowing much about the case. But it would be just too great a coincidence for independent witnesses to make up the hemlock story. The hemlock story is surprising information, and it seems that the more surprising the information is, the more confident we may be that the information is true.

In Chapter 3, we commented on C. I. Lewis's claim that when many independent witnesses tell us the same story, we can be virtually certain that they are reliable witnesses, since if the story were not true, it is just one out of so many false stories that could have been told. Now if the witnesses tell us that the victim was shot, then this is not simply one out of so many false stories. It is the sort of thing that someone is likely to say if she does not know, or does not wish to convey, any details about the person missing in action, but feels compelled to say *something*. However, there are many stories that are as odd as the hemlock story, and it would be just too strange for all of the witnesses to independently make up that *same* story. The hemlock story, repeated by independent witnesses, is just too odd not to be true. In Lewis's words: '[O]n any other hypothesis than that of truth-telling, this agreement is very unlikely' (1946: 346).

The same observation is made by P. O. Ekelöf (1983: 22). He proposes a formula to calculate the rational degree of confidence after concurring reports. However, he concedes that this formula does not yield the right results when two witnesses report something 'quite extraordinary'. In this case, each report taken separately may have very little evidential value, but both reports taken together raise our degree of confidence in what is reported far beyond the value that the Ekelöf formula yields. Ekelöf does not give us any further hint about how one should assess concurring witness reports containing such information. We take on that task in this chapter.

5.2. TESTIMONIES FROM INDEPENDENT WITNESSES

The more odd a witness report is, the more unlikely it is that an unreliable witness would hit on it. An unreliable witness does not look at the world, but just makes up a story. Such a story could consist in answering a simple yes-or-no question, but it could also consist in answering an open-ended question, such as what happened to the person missing in action. In the models that we constructed in Chapter 3, the chance that an unreliable witness hits on a report is characterized by the randomization parameter $a = \mathbf{P}(\text{REP}_i|\text{HYP}, \neg\text{REL}_i) = \mathbf{P}(\text{REP}_i|\neg\text{HYP}, \neg\text{REL}_i)$ for witnesses $i = 1, \ldots, n$. And generally, the more odd the story is, the lower the value the randomization parameter should have. We stipulated that the witnesses are equally reliable and calculated the posterior probability that a single witnesses is reliable and the posterior probability that the story the witnesses tell is true after receiving n reports to the effect that the story is indeed true. In Appendix E.1, we generalize this result for witnesses who may not be equally reliable,

$$\text{(5.1)} \qquad \mathbf{P}^{*(n)}(\text{HYP}) = \mathbf{P}(\text{HYP}|\text{REP}_1, \ldots, \text{REP}_n) = \frac{h}{h + \bar{h}\prod_{i=1}^{n} x_i}$$

$$\text{with the likelihood ratios} \quad x_i = \frac{\mathbf{P}(\text{REP}_i|\neg\text{HYP})}{\mathbf{P}(\text{REP}_i|\text{HYP})} = \frac{a\bar{\rho}_i}{\rho_i + a\bar{\rho}_i}$$

for independent witnesses $i = 1, \ldots, n$.

In Appendix E.2, we calculate the posterior probability that a single witness i is reliable:

$$(5.2) \qquad P^{*(n)}(\text{REL}_i) = P(\text{REL}_i | \text{REP}_1, \ldots, \text{REP}_n) = \frac{h\bar{x}_i}{h + \bar{h}\prod_{j=1}^{n} x_j} = \bar{x}_i P^{*(n)}(\text{HYP}).$$

The partial derivative of $P^{*(n)}(\text{HYP})$ with respect to the randomization parameter a yields the rate of increase or decrease in $P^{*(n)}(\text{HYP})$ as a increases. This partial derivative is always negative (see Appendix E.3), which indicates that $P^{*(n)}(\text{HYP})$ decreases as a increases. And we get a similar result for $P^{*(n)}(\text{REL}_i)$:

$$(5.3) \qquad \frac{\partial P^{*(n)}(\text{HYP})}{\partial a} < 0, \text{ and } \quad \frac{\partial P^{*(n)}(\text{REL}_i)}{\partial a} < 0.$$

Hence, the less likely it is that unreliable witnesses would report some story or other, the more confident we may be that the story is true and that an arbitrary witness is reliable, ceteris paribus.

But we need to be extremely careful about the interpretation of the ceteris paribus clause here. Presently, we are assuming that *we*—i.e. the police inspector, the judge, or the members of a jury—assign equal prior probability values h to the hypothesis that the person missing in action was shot as we would to the hypothesis that he was made to drink hemlock. Yet we find it much less likely that an unreliable witness would concoct a story about being made to drink hemlock than she would concoct a story about being shot. But the assumption of equal prior probabilities is rather unrealistic. Typically we share some background knowledge with the witnesses. If we believe that *unreliable witnesses* are unlikely to make up a particular story, this is so because *we* think that it is a rather implausible story. So it seems that the parameters a and h will usually co-vary. An odd story is a story that we neither expect to be true nor expect the witnesses to make up.[1]

To illustrate a case in which a and h do not only co-vary, but are identical, let us go back to Noah's predicament in our Introduction. Noah has no clue as to which of his sons so shamelessly caught a glimpse of him naked. He decides to ask two of his friends who have the role of independent witnesses. They may or may not know anything, but they are always eager to please. If they do not know anything, they will simply randomize between the suspects and call a name. So if Noah has n children, then the chance a that an unreliable witness picks Ham is $1/n$. Since Noah has no clue which of his children is the culprit, the prior probability h that it is Ham who is the culprit is also $1/n$. Certainly Noah would feel more confident if he had had a dozen children than if he had had two children and both of his friends picked Ham. With two children, there is a very real chance that two unreliable witnesses just happened to pick out an innocent Ham, whereas it really would be a coincidence that they had both picked an innocent Ham out of a dozen children. But would Noah be more confident when there are two versus three children? Three versus four? There are two considerations at work here. With fewer children, the prior probability h that Ham is the culprit is higher so that we would expect the posterior probability to be higher as well. But with more children, the randomization parameter is lower as well, so that we would expect the posterior probability that Ham is the culprit to be higher. So how do these considerations play out against each other? And how do they play out against each other in cases with more independent witnesses providing concurring information?

5.3. LOWER PRIORS, HIGHER POSTERIORS?

Olsson (2002*a*) ascribes the claim to Jonathan L. Cohen (1977) that the lower the prior probability of the information provided by the witnesses is, the higher the posterior probability is that the information is true.

[1] Granted, the parameters a and h do not *need* to co-vary. For instance, it could well be that we genuinely think that every suspect has an equal chance of being the criminal, while we do not doubt that unreliable witnesses are much more likely to pick Scarface over

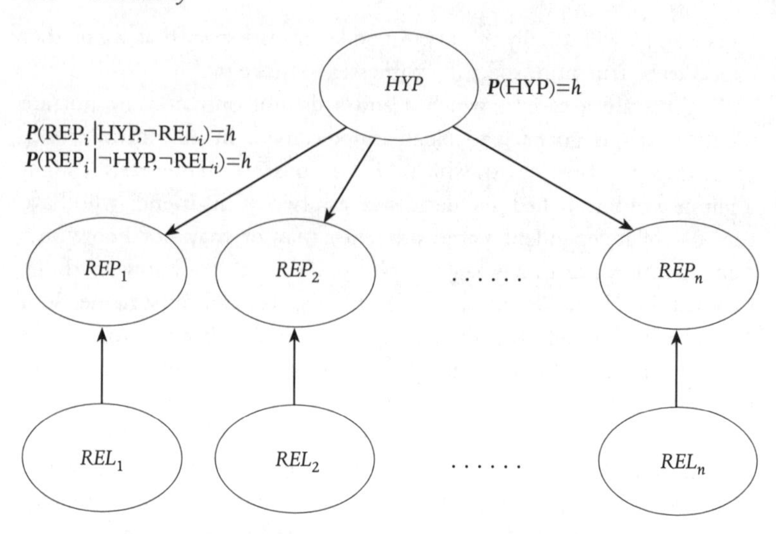

Fig. 5.1 The Bayesian Network for multiple reports about a single fact with independent witnesses and reliability defined endogenously and the value of the randomization parameter *a* set at *h*

Olsson in effect constructs a model in which the reliability of the witnesses functions as an endogenous variable and sets the value of the randomization parameter *a* equal to the prior probability value for the hypothesis *h*. Bovens *et al.* (2002) have argued that Olsson misinterprets Cohen's claim and that his model does not respect the independence assumption that Cohen requires. We will not rehash the details of this discussion. But we will build on Olsson's idea and investigate what happens when the value of the randomization parameter *a* is set equal to *h* in the Bayesian Network for independent witnesses represented in Figure 3.5 of Chapter 3. We present this Bayesian Network in Figure 5.1.

In (5.1), we calculated the posterior probability of the information that we received from independent witnesses with reliability modelled as an endogenous variable. We now substitute *h* for *a* in this formula:

$$(5.4) \qquad P^{*(n)}(\mathrm{HYP}) = P(\mathrm{HYP}|\mathrm{REP}_1, \ldots, \mathrm{REP}_n) = \frac{h}{h + \bar{h}\prod_{i=1}^{n} x_i}$$

Peacenik on the basis of their looks alone. In that case, the partial derivatives in (5.3) give us an accurate assessment. We would be more impressed with the evidence if *n* independent witnesses had all picked Peacenik than if *n* independent witnesses had all picked Scarface.

with the likelihood ratios $x_i = \frac{h\bar{\rho}_i}{\rho_i + h\bar{\rho}_i}$ for independent witnesses $i = 1, \ldots, n$.

Let us first consider the case in which Noah consults two of his friends ($n = 2$). The formula in (5.4) simplifies to:

$$P^{*(2)}(\text{HYP}) = \frac{(h + \bar{h}\rho_1)(h + \bar{h}\rho_2)}{h + \bar{h}\rho_1\rho_2}. \tag{5.5}$$

Is the posterior probability that Ham is the culprit after receiving two reports to this effect greater when Noah has two sons ($h = 1/2$), than when he has three sons ($h = 1/3$), or ten sons ($h = 1/10$)? In Figure 5.2, we plot $P^{*(2)}(\text{HYP})$ as a function of h for independent witnesses with the same reliability for $\rho_1 = \rho_2 = .10, .30, .60$. The general shape of these curves is not surprising. When the prior probability h of an item of information approaches 0, it becomes increasingly unlikely that we would have received this information from unreliable witnesses. Hence, if we *did* receive this information, then it is likely to be from reliable witnesses, which raises the posterior probability $P^{*(2)}(\text{HYP})$. When the prior probability h of an item of information approaches 1, then the posterior probability $P^{*(2)}(\text{HYP})$ approaches 1 for confirming

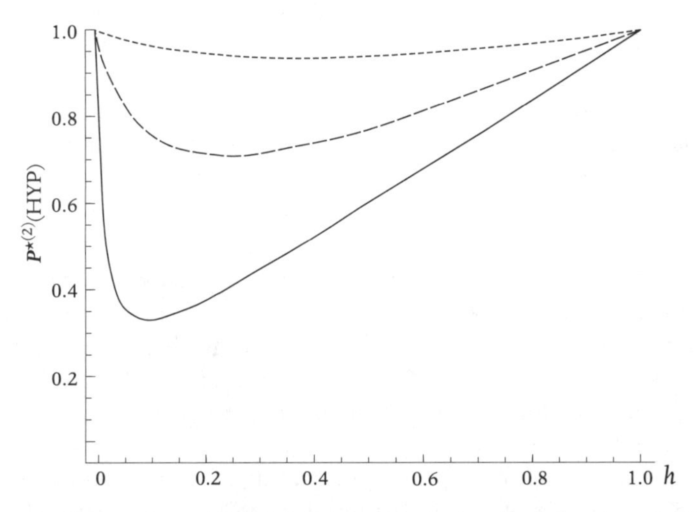

FIG. 5.2 The posterior probability of the hypothesis as a function of the prior probability of the hypothesis when the witnesses are independent and $\rho_i = .10$ (full line), .30 (dashed line) and .60 (dotted line).

reports to the effect that the information is true. When we are ambivalent about some item of information, i.e. when h is not approaching 0 or 1, then the posterior probability that the information is true will very much depend on how reliable we take the witnesses to be. But note that when we take the witnesses to be very unreliable, i.e. $\rho_1 = \rho_2 =: \rho = .10$, then the posterior probability that Ham is the culprit is greater for 2 than for 3 children and greater for 3 than for 10 children. For $\rho = .30$, and similarly for $\rho = .60$, the posterior probability that Ham is the culprit is greater for 10 than for 2 children and greater for 2 than for 3 children. For ρ approaching 1 (not depicted), the posterior probability is greater for 10 than for 3 children and greater for 3 than for 2 children. Furthermore, as Jørgen Hilden (2002) observed, one can prove (see Appendix E.4) the more general claim that for $\rho \geq \frac{1}{2}\sqrt{2} \approx .707$ the posterior probability that Ham is the culprit is greater for $n + 1$ children than for n children for $n \geq 2$.

The reason for this shift is that the function has different minima for different values of ρ. For values to the left of this minimum, the greater the value of h is, the lower the value of $\boldsymbol{P}^{*(2)}(\text{HYP})$ is. In this range of values of h, the more surprising the information is, the more confident we will be that the information is true. For values to the right of this minimum, the greater the value of h is, the greater the value of $\boldsymbol{P}^{*(2)}(\text{HYP})$ is. In this range of values of h, the more surprising the information is, the less confident we will be that the information is true. What we notice is that for lower values of ρ, the minimum shifts towards the left. The progression of the curves in Figure 5.2 for lower values of ρ is easy enough to understand if we consider the limiting values of ρ. As ρ approaches 0, $\boldsymbol{P}^{*(2)}(\text{HYP})$ will approach h. A witness who is known to consult a randomizing device before pointing to Ham is of no consequence in changing Noah's prior beliefs. As ρ approaches 1, $\boldsymbol{P}^{*(2)}(\text{HYP})$ will approach 1. If God had pointed to Ham, Noah would have had no choice but to accept his verdict, no matter how many children he had. So the graph progresses from a diagonal curve for $\boldsymbol{P}^{*(2)}(\text{HYP}) = h$ to a flat curve at $\boldsymbol{P}^{*(2)}(\text{HYP}) = 1$ as ρ increases from 0 to 1. And this is precisely what we see in Figure 5.2.

If we set the reliability parameter at a particular value, then there is a range of values of h which permit us to make the following claim: *The lower the prior probability that the hypothesis is true, the greater the posterior probability that the hypothesis is true after receiving two witness* *reports to this effect*. The width of this range of values for h is $(0, h_{\min}]$, in which the value of $h_{\min}$ is the minimum of the function $\boldsymbol{P}^{*(2)}(\text{HYP})$. To find the minima for different values of ρ, we calculate the partial derivative with respect to h,

$$\frac{\partial \boldsymbol{P}^{*(2)}(\text{HYP})}{\partial h} = \bar{\rho}_1\bar{\rho}_2 \frac{h^2 - \bar{h}^2 \rho_1\rho_2}{(h + \bar{h}\rho_1\rho_2)^2}, \tag{5.6}$$

set this partial derivative to 0 and solve for h.[2] The solution for $h \in (0, 1)$ is

$$h_{\min} = \frac{\sqrt{(\rho_1\rho_2)}}{1 + \sqrt{(\rho_1\rho_2)}}. \tag{5.7}$$

We plot this function in Figure 5.3 letting $\rho := \rho_1 = \rho_2$. When the witnesses are independent, the claim that a lower prior probability entails a greater posterior probability holds true for all the values of ρ and h underneath the curve.

So far we have conducted our analysis within the framework of *decision-making under risk*. Our models presuppose precise probabilities that the witnesses are reliable. But in reality we often are not capable of assessing these probabilities. Rather, the situation is one of *decision-making under uncertainty*: We have no clue whatsoever about the chance that the witness is reliable or not. To represent such ignorance, let ρ be a continuous variable whose values range from 0 (for certainty that the witnesses are fully unreliable) to 1 (for certainty that the witnesses are fully reliable). We represent our ignorance of the reliability of witnesses 1 and 2 by assuming that ρ_1 and ρ_2 are both uniformly distributed over the interval [0, 1].[3] If we have no clue whether the

[2] To assure ourselves that this solution yields a minimum rather than a maximum over $\rho_1, \rho_2 \in (0, 1)$, observe that the second derivative of $\boldsymbol{P}^{*(2)}$ (HYP) with respect to h is

$$\frac{\partial^2 \boldsymbol{P}^{*(2)}(\text{HYP})}{\partial^2 h^2} = \frac{2\rho_1\bar{\rho}_1\rho_2\bar{\rho}_2}{(h + \bar{h}\rho_1\rho_2)^3} > 0 \text{ for } h, \rho_1, \rho_2 \in (0, 1).$$

[3] This is an appeal to the Bayes–Laplace Postulate, which has elicited much discussion in the Bayesian literature (e.g. Bernardo and Smith (2000: Sect. 5.6.2) and Robert (2001: Sect. 3.5)). We take the uniform distribution over the prior probability of the reliability of a witness to be *one* representation and not *the one and only* representation of our ignorance with respect to the chance that a witness is reliable. Standard objections to the Principle of

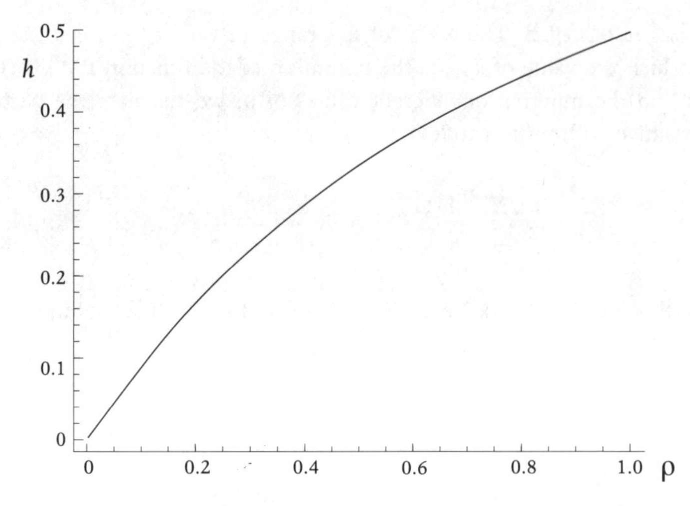

FIG. 5.3 The space under the curve represents all points in which the posterior probability of the hypothesis is a decreasing function of the prior probability of the hypothesis when the witnesses are independent

witnesses are reliable in the sense that we take any value of ρ_i to be equally probable (for $i = 1, 2$), then for what values of h can we say that a lower prior probability entails a greater posterior probability? We calculate the expected value of $\boldsymbol{P}^{*(2)}$(HYP) for each value of h:

$$(5.8) \qquad < \boldsymbol{P}^{*(2)}(\text{HYP}) >= \int_0^1 \int_0^1 \frac{(h + \bar{h}\rho_1)(h + \bar{h}\rho_2)}{h + \bar{h}\rho_1\rho_2} d\rho_1 d\rho_2.$$

The analytical expression of this function is quite complex. We have plotted $< \boldsymbol{P}^{*(2)}(\text{HYP}) >$ in Figure 5.4. To find the smallest expected value of $\boldsymbol{P}^{*(2)}$(HYP), we take the derivative of $< \boldsymbol{P}^{*(2)}(\text{HYP}) >$ with respect to h, set this derivative equal to 0 and solve for $h \in (0,\ 1)$. The result is this:

$$(5.9) \qquad h_{\min} \approx .199.$$

Indifference, such as Bertram's paradox, do not apply here (*cf.* Olsson (2002*c*: 570–1)). We are interested in probabilities and not, say, in odds. Hence the probability that a witness is reliable is the natural parameter of interest, and there is no reason to insist on the invariance of the uniform distribution under alternative parameterizations.

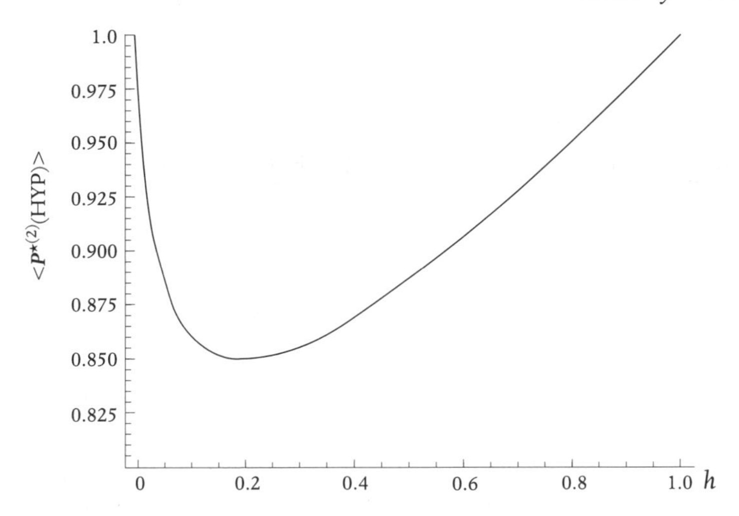

FIG. 5.4 The posterior probability of the hypothesis after concurring reports from two independent witnesses as a function of the prior probability of the hypothesis under conditions of decision-making under uncertainty

Notice that $1/5 \approx .199$. Hence, if Noah has no clue whatsoever whether his friends are reliable or not, then his minimum confidence that Ham is the culprit, given two reports to that effect, would occur if he had five children. He would be more confident that Ham is the culprit if he had six rather than five children, but he would also be more confident if he had four rather than five children.

5.4. GENERALIZING TO MANY WITNESSES

What happens when we increase the number of witnesses? It is no surprise that for a fixed value of the reliability of the witnesses the posterior probability that the information is true when all witnesses agree will be higher for more witnesses. In Figure 5.5, we have plotted the posterior probability $P^{*(n)}$ as presented in (5.4) with $n = 2,\ 5$, and 11 witnesses and $\rho_i = .10$ for $i = 1, \ldots, n$. As we might expect, as the number of witnesses increases, the curves start resembling the curves for two witnesses with higher reliability. Receiving information from more witnesses who are less likely to be reliable yields a similar degree of confidence as receiving information from fewer witnesses who are more likely to

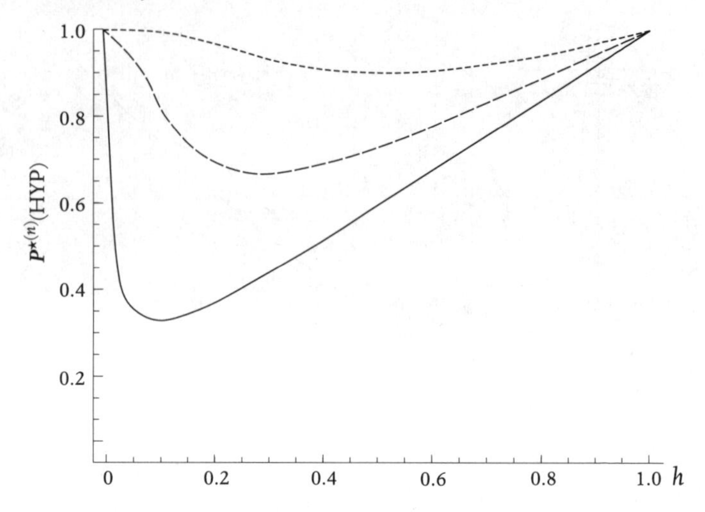

Fig. 5.5 The posterior probability of the hypothesis after concurring reports from $n = 2$ (full line), 5 (dashed line), and 11 (dotted line) independent witnesses as a function of the prior probability of the hypothesis for $\rho = .10$

be reliable. So we get the same reversal of Noah's confidence that we noticed before as we increased the reliability of the two independent witnesses. For 2 witnesses, Noah's confidence that Ham is the culprit is greater if he has 2 rather than 3 children and is greater if he has 3 rather than 10 children. For 5 witnesses, his confidence is greater for 10 rather than 3 children and for 3 rather than 2 children. And for 11 witnesses, his confidence is greater for 2 rather than 3 children and for 3 rather than 10 children.

We show in Appendix E.5 that if all ρ_is are equal and set at ρ, then the minima for these functions can be expressed in a simple form:

$$h_{\min} = \frac{(n-1)\rho}{1+(n-1)\rho}. \tag{5.10}$$

We have plotted $h_{\min}$ as a function of n in Figure 5.6 for three different values of ρ. Notice that when ρ is .10, the claim that a lower prior probability entails a greater posterior probability is true for *all* values of $h \in (0, .50)$ if and only if there are 11 or more concurring reports. For witnesses who are less likely to be reliable, it takes more concurring reports for the claim that a lower prior probability entails a greater posterior probability to hold over a wide range of values of h.

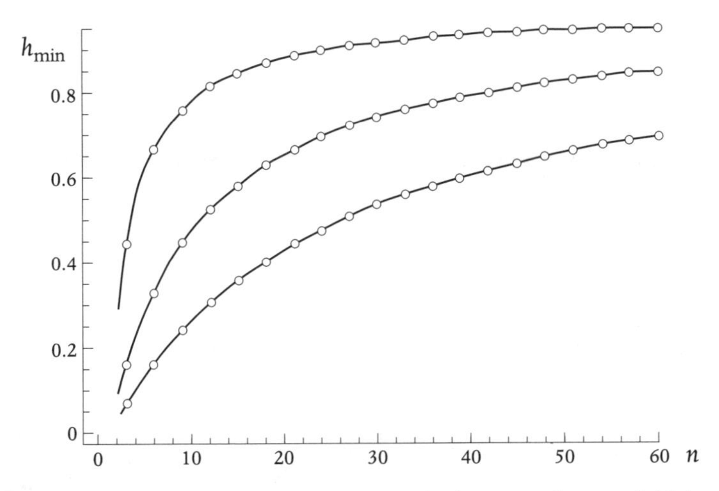

FIG. 5.6 A smaller prior entails a greater posterior for the ranges of prior probabilities of the hypothesis $(0, h_{min})$ under decision-making under risk when there are n independent witnesses with $\rho_i = .04$ (bottom curve), $\rho_i = .1$ (middle curve) and $\rho_i = .4$ (top curve) for $i = 1, \ldots, n$

For decision-making under uncertainty, as interpreted earlier, we have plotted h_{min} as a function of the number of witnesses n in Figure 5.7. In Appendix E.6, we show how this plot can be obtained numerically. Notice that the claim that a lower prior probability entails a greater posterior probability is true for *all* values of $h \in (0, .50)$ when there are five or more concurring reports. As the number of witnesses n providing concurring reports increases, the range of values for which the claim that a lower prior probability entails a greater posterior probability holds broadens, and goes to $(0, 1)$ as n goes to infinity.

5.5. SHOPPING FOR CONSUMER PRODUCTS

The situation of Noah rests on two very special assumptions, viz. (i) all the alternative hypotheses are equally probable and (ii) an unreliable witness acts as if the best she can do in providing an answer is to randomize over the possible options. This story can be made plausible for witnesses attempting to identify a culprit from among a line-up of suspects. It may well be that it seems equally probable that each suspect is the criminal. And it may well be that fully unreliable witnesses act as if

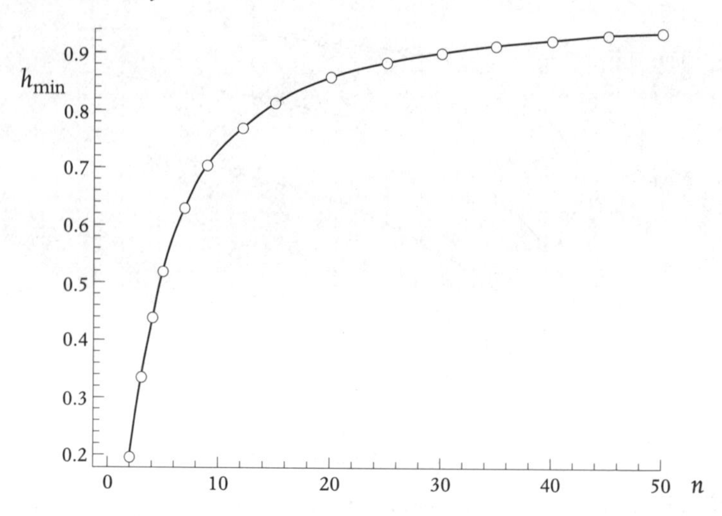

FIG. 5.7 A smaller prior entails a greater posterior for the ranges of prior probabilities of the hypothesis $(0, h_{min})$ when there are n independent witnesses under conditions of decision-making under uncertainty

they are randomizers with a $1/n$ chance of picking any particular suspect. But, presumably, concurrence in witness reports concerning a 'quite extraordinary' option also constitutes strong evidence, even if unreliable witnesses do not have a uniform distribution over the alternatives.

Suppose, for instance, that you are shopping around for some specialized consumer product, say a bread machine. There is *one* more well-known brand and a range of less well-known brands on the market. It is common knowledge that better products tend to come from more well-known brands, but there are exceptions to this rule. Hence, the prior probability that the better product comes from the more well-known brand exceeds the prior probability that it comes from one of the less well-known brands. You now solicit advice from independent sources. It is not clear to you who among them really knows anything more than you do about the consumer product in question. Compare two cases. In case one, all of your sources recommend the more well-known brand. In case two, all of your sources recommend the same very obscure brand. In which case is your degree of confidence greater that the recommended product is the better product? Presumably 'too-odd-not-to-be-true' reasoning should carry some weight here as well. Sometimes, we *will* find ourselves more convinced by the concurring recommendations for the less well-known brand, while the recommendations for the better-known brand gives us no reason to believe that our informers know any more than we do. A general model of 'too-odd-not-to-be-true' reasoning should also be capable of handling cases where there is not a uniform distribution over the possible alternatives. But it is not plausible in this case to stipulate that $a = \mathbf{P}(\text{REP}|\text{HYP}, \neg\text{REL}) = \mathbf{P}(\text{REP}|\neg\text{HYP}, \neg\text{REL}) = \mathbf{P}(\text{HYP}) = h$. What can be done to model this situation?

What is an unreliable witness in this case? One possibility is that it is a person who knows precisely what we know. Such a person will just pick the more probable alternative, i.e. the product from the better-known brand. Another possibility is that it is a person who is completely ignorant. Such a person will just randomize over the n alternative products with probability $1/n$. So let us model the situation by altering the variables REL_i in the Bayesian Network in Figure 3.7 (in Chapter 3) to make them ternary propositional variables. Their values are REL^j_E (witness j is an expert), REL^j_B, (witness j has the same beliefs as we do), and REL^j_I (witness j is fully ignorant). Furthermore, let us suppose that there is one well-known brand and $n-1$ less well-known brands. So the variables REP can take on n values, viz. REP^j_{wk} (witness j names the well-known brand) and REP^j_{lwk-i} (witness j names a less well-known brand for $i = 1, \ldots, n-1$). The variable HYP can now take on n values, viz. HYP_{wk} (the well-known brand offers the best product), and HYP_{lwk-i} (less well-known brand i offers the best product for $i = 1, \ldots, n-1$). We specify the relevant probabilities in Table 5.1.

Let us now calculate the posterior probability that the well-known brand indeed offers the better product, given two reports to this effect, and the posterior probability that one of the $n-1$ less well-known brands offers the better product, given two reports to this effect:

$$(5.11) \quad \mathbf{P}^{*(2)}(\text{HYP}_{wk}) = \mathbf{P}\left(\text{HYP}_{wk}|\text{REP}^1_{wk}, \text{REP}^2_{wk}\right) = \frac{h_{wk}}{h_{wk} + \bar{h}_{wk}x^2_{wk}} \text{ with } x_{wk} = \frac{n(\overline{\rho}_I - \rho_E) + \rho_I}{n\overline{\rho}_I + \rho_I}.$$

TABLE 5.1 *The marginal and conditional probabilities for the Bayesian Network in Figure 3.7 as a model for 'too-odd-not-to-be-true' reasoning for* $\forall j = 1, \dots n-1$

$P(\text{REL}_E) = \rho_E$; $P(\text{REL}_B) = \rho_B$; $P(\text{REL}_I) = \rho_I$	
$P(\text{HYP}_{wk}) = h_{wk}$	$P(\text{HYP}_{lwki}) = h_{lwk}$
$P(\text{REP}^j_{wk} \mid \text{HYP}_{wk}, \text{REL}^j_E) = 1$	$P(\text{REP}^j_{lwk\text{-}i} \mid \text{HYP}_{wk}, \text{REL}^j_E) = 0$
$P(\text{REP}^j_{wk} \mid \text{HYP}_{lwk\text{-}i}, \text{REL}^j_E) = 0$	$P(\text{REP}^j_{lwk\text{-}i} \mid \text{HYP}_{lwk\text{-}i}, \text{REL}^j_E) = 1$
$P(\text{REP}^j_{wk} \mid \text{HYP}_{wk}, \text{REL}^j_B) = 1$	$P(\text{REP}^j_{lwk\text{-}i} \mid \text{HYP}_{wk}, \text{REL}^j_B) = 0$
$P(\text{REP}^j_{wk} \mid \text{HYP}_{lwk\text{-}i}, \text{REL}^j_B) = 1$	$P(\text{REP}^j_{lwk\text{-}i} \mid \text{HYP}_{lwk\text{-}i}, \text{REL}^j_B) = 0$
$P(\text{REP}^j_{wk} \mid \text{HYP}_{wk}, \text{REL}^j_I) = 1/n$	$P(\text{REP}^j_{lwk\text{-}i} \mid \text{HYP}_{wk}, \text{REL}^j_I) = 1/n$
$P(\text{REP}^j_{wk} \mid \text{HYP}_{1wk\text{-}i}, \text{REL}^j_I) = 1/n$	$P(\text{REP}^j_{1wk\text{-}i} \mid \text{HYP}_{1wk\text{-}i}, \text{REL}^j_I) = 1/n$

(5.12)
$$P^{*(2)}(\text{HYP}_{1wk}) = P(\text{HYP}_{1wk} \mid \text{REP}^1_{1wk}, \text{REP}^2_{1wk}) = \frac{h_{lwk}}{h_{lwk} + \bar{h}_{lwk} x^2_{lwk}} \text{ with } x_{lwk} = \frac{\rho_I}{\rho_I + n\rho_E}.$$

Let us look at a numerical example to see what these equations tell us. We are shopping for a bread machine and have a choice between the famous brand *Hotstuff* and the less well-known brands *Cheapo*, *Shoddy*, and *Trashy*. Clearly, $h_{wk} > h_{lwk}$ so that $h_{wk} > 1/n$. In this case, $h_{wk} > 1/4$. Let the prior probabilities that one of the less well-known brands carries the better products be equal, so that $h_{lwk} = (1 - h_{wk})/(n-1)$. In this case,

(5.13)
$$h_{lwk} = (1 - h_{wk})/3.$$

Furthermore let us assume that there is a one-third chance that a witness is an expert, a one-third chance that a witness knows as much as we do and a one-third chance that a witness is completely ignorant:

(5.14)
$$\rho_E = \rho_B = \rho_I = 1/3.$$

We insert (5.13) and (5.14) into (5.11) and (5.12) and obtain

(5.15)
$$P^{*(2)}(\text{HYP}_{wk}) = \frac{81 h_{wk}}{56 h_{wk} + 25} \text{ and}$$
$$P^{*(2)}(\text{HYP}_{lwk}) = \frac{25 \bar{h}_{wk}}{24 \bar{h}_{wk} + 3}.$$

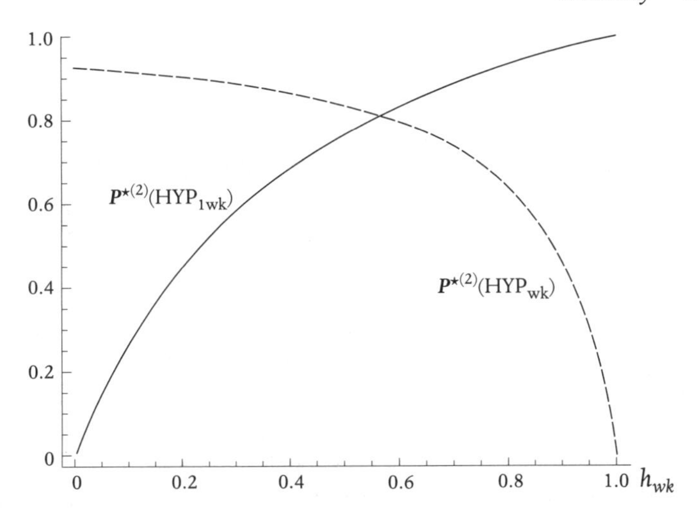

FIG. 5.8 The posterior probability that one of three less well-known brands offers the better product after two reports to this effect and the posterior probability that the well-known brand offers the better product after two reports to this effect as a function of the prior probability that the well-known brand offers the better product

These functions are plotted in Figure 5.8. Let h_{wk} equal .55 so that $h_{lwk} = .15$. In this graph, $P^{*(2)}(\text{HYP}_{\text{wk}}) < P^{*(2)}(\text{HYP}_{\text{lwk}})$. So, the posterior probability that *Hotstuff* offers the better bread machine, after getting two reports to this effect, is smaller than the posterior probability that *Cheapo* offers the better bread machine, after getting two reports to this effect. The two curves cross at $h_{wk} \approx .57$. Hence, for $h_{wk} \in (.25, .57)$, the posterior probability that *Cheapo* offers the better product is greater than the posterior probability that *Hotstuff* offers the better product. This result can easily be generalized to one well-known brand and $n - 1$ less well-known brands. In this case the interval in question becomes $(1/n, h_{cross}(n))$. As the number of less well-known brands goes to infinity, $h_{cross}(n)$ converges to 1 (proof omitted) and $1/n$ converges to 0.

It is easy to see why this is the case. With lots of brands on the market, it becomes very unlikely that two fully ignorant witnesses would happen to name the same less well-known brand. Hence it becomes very likely that such information comes from expert witnesses. So the posterior probability that the less well-known brand offers the better product is higher than the posterior probability that the well-known brand offers the better product. And this is the case for a broad range of values of the prior probability that the well-known brand offers the better product.

Epilogue

We have not tried to present an overview of current work in Bayesian epistemology. Rather, we have picked some epistemological themes in which probabilistic modelling proves to be fruitful. This methodology is commonplace in the natural and social sciences, but it is not a standard philosophical tool, notwithstanding the attention that probability has recently received in philosophical arguments. We believe that there are many issues in philosophy that could benefit from this methodology. We urge the reader to use this book in the first place as a methodological toolbox with instruments that are often not part of a standard philosophical curriculum. Over the years we have encountered many philosophical problems that have led us to draw up blueprints of models. Some of these blueprints were promising, whereas others were not. Let us look at some of the virtues and vices of probabilistic modelling in philosophy.

A model helps us draw out and clarify intuitions. It plays the Socratic role of midwife, and gives birth to what we already knew but were not able to articulate in a clear and perspicuous manner. This is what we are doing when we make use of Bayesian Networks to draw distinctions between different types of relatively unreliable witnesses or to present an interpretation of what constitutes variety of evidence.

Models also prove helpful when we are faced with philosophical questions to which our intuitions provide no answers whatsoever. Our intuitions just give out or present us with some *on the one hand—on the other hand* story. This is the time when we hear Leibniz's *dictum*: 'Let us calculate, Sir.' For instance, the coherence of an information set seems to have something to do with our willingness to believe the information, but our intuitions do not go much further in spelling out the role of this determinant for our degree of confidence. Also, our intuitions provide no clear answer to the question of whether a hypothesis receives stronger confirmation from a supporting item of evidence when the hypothesis and the auxiliary theories are independent or when they are dependent. And finally, consider 'too-odd-not-to-be-true' reasoning. On the one hand, we don't tend to believe highly unlikely information, even if it is reported by more than one source. But on the other hand, sometimes we are prone to believe information that is just too odd not to be true—there is no way that our sources could have concocted such a story independently of one another. A carefully constructed model tells us under what conditions one rather than the other consideration is weightier.

A good-making feature of modelling is that it implies some conclusions that seem to be counter-intuitive at first. Our claim that there cannot exist a coherence ordering over information sets falls into this category, as does our claim that there are plausible interpretations of the variety-of-evidence thesis that make the thesis come out false in some cases. The model is used to support some challenging claims that engage our philosophical curiosity.

But modelling can be a thankless job as well. Philosophers are not satisfied when they see some computer-generated graph with phase curves. We want to know *why* it is that the phenomenon is determined by the values of the parameters in this way or other. Showing the equation does not satisfy this desire for understanding. Somehow we want to hear a story that has some intuitive appeal. But if we are able to tell this story too well, then it is not clear what the model has left to offer. If we don't tell the story well enough, then the model fails to help us understand the issue. One needs to work this middle ground and show how our pre-theoretical understanding gives out at a certain point, how the model provides a precise answer to the question at hand, and then make at least some suggestive remarks in the direction of the results of the model to improve our intuitive understanding of the issue. These remarks should be resolute rather than half-hearted, but they should remain suggestive remarks. If we are able to do more than make suggestive remarks, then the model becomes gratuitous. And indeed, many models become gratuitous *post hoc*. Once one starts thinking about what the model tells us, one realizes that these very same results could also have been reached without doing the mathematical analysis—just by thinking clearly about the issue in question. The waste-paper basket is the only proper place for such results.

There is another trap that it is easy for model builders to fall into. You may have heard the academic joke about the economist among the shipwreck survivors on a desert island. A carton of cans from the sunken ship drifts ashore. The economist in the group steps forward and declares 'Suppose that we had a can opener...' Indeed, it is easy to distort the problem under consideration by making implausible assumptions that make the modelling easier. And it is often all too easy to get carried away by the formulae. Not every partial derivative is meaningful—no matter how pretty the result is.

Perhaps some of our work displays the vices of modelling. But then, philosophical tastes are varied, and we hope that our readers have found something here that sparked their interests. And foremost, we hope that we have provided inspiration to take up similar tools to address other philosophical problems.

Appendix

A. PROOF OF CHAPTER I

A.1. *Eq. (1.5)*

We first prove the equation for $n = 2$. We apply Bayes Theorem:

$$\begin{aligned} \boldsymbol{P}^*(\mathrm{R}_1, \mathrm{R}_2) &= \boldsymbol{P}(\mathrm{R}_1, \mathrm{R}_2 | \mathrm{REPR}_1, \mathrm{REPR}_2) \\ &= \frac{\boldsymbol{P}(\mathrm{REPR}_1, \mathrm{REPR}_2 | \mathrm{R}_1, \mathrm{R}_2)\, \boldsymbol{P}(\mathrm{R}_1, \mathrm{R}_2)}{\boldsymbol{P}(\mathrm{REPR}_1, \mathrm{REPR}_2)}. \end{aligned} \tag{A.1}$$

By the chain rule and the independence assumptions in (1.3), we obtain:

$$\begin{aligned} \boldsymbol{P}(\mathrm{REPR}_1, \mathrm{REPR}_2 | \mathrm{R}_1, \mathrm{R}_2) &= \boldsymbol{P}(\mathrm{REPR}_1 | \mathrm{REPR}_2, \mathrm{R}_1, \mathrm{R}_2)\, \boldsymbol{P}(\mathrm{REPR}_2 | \mathrm{R}_1, \mathrm{R}_2) \\ &= \boldsymbol{P}(\mathrm{REPR}_1 | \mathrm{R}_1)\, \boldsymbol{P}(\mathrm{REPR}_2 | \mathrm{R}_2). \end{aligned} \tag{A.2}$$

We expand the denominator of (A.1):

$$\begin{aligned} \boldsymbol{P}(\mathrm{REPR}_1, \mathrm{REPR}_2) = {} & \boldsymbol{P}(\mathrm{REPR}_1, \mathrm{REPR}_2 | \mathrm{R}_1, \mathrm{R}_2)\, \boldsymbol{P}(\mathrm{R}_1, \mathrm{R}_2) + \\ & \boldsymbol{P}(\mathrm{REPR}_1, \mathrm{REPR}_2 | \neg\mathrm{R}_1, \mathrm{R}_2)\, \boldsymbol{P}(\neg\mathrm{R}_1, \mathrm{R}_2) + \\ & \boldsymbol{P}(\mathrm{REPR}_1, \mathrm{REPR}_2 | \mathrm{R}_1, \neg\mathrm{R}_2)\, \boldsymbol{P}(\mathrm{R}_1, \neg\mathrm{R}_2) + \\ & \boldsymbol{P}(\mathrm{REPR}_1, \mathrm{REPR}_2 | \neg\mathrm{R}_1, \neg\mathrm{R}_2)\, \boldsymbol{P}(\neg\mathrm{R}_1, \neg\mathrm{R}_2). \end{aligned}$$

Following the same reasoning as in (A.2):

$$\boldsymbol{P}(\mathrm{REPR}_1, \mathrm{REPR}_2 | \neg\mathrm{R}_1, \mathrm{R}_2) = \boldsymbol{P}(\mathrm{REPR}_1 | \neg\mathrm{R}_1)\, \boldsymbol{P}(\mathrm{REPR}_2 | \mathrm{R}_2), \tag{A.3}$$

$$\boldsymbol{P}(\mathrm{REPR}_1, \mathrm{REPR}_2 | \mathrm{R}_1, \neg\mathrm{R}_2) = \boldsymbol{P}(\mathrm{REPR}_1 | \mathrm{R}_1)\, \boldsymbol{P}(\mathrm{REPR}_2 | \neg\mathrm{R}_2), \tag{A.4}$$

$$\boldsymbol{P}(\mathrm{REPR}_1, \mathrm{REPR}_2 | \neg\mathrm{R}_1, \neg\mathrm{R}_2) = \boldsymbol{P}(\mathrm{REPR}_1 | \neg\mathrm{R}_1)\, \boldsymbol{P}(\mathrm{REPR}_2 | \neg\mathrm{R}_2). \tag{A.5}$$

We substitute a_0, a_1, a_2, p, q, and r:

$$\begin{aligned}
\boldsymbol{P}^*(\mathrm{R}_1, \mathrm{R}_2) &= \frac{p^2 a_0}{p^2 a_0 + pq\ \boldsymbol{P}(\neg\mathrm{R}_1, \mathrm{R}_2) + pq\ \boldsymbol{P}(\mathrm{R}_1, \neg\mathrm{R}_2) + q^2 a_2} \\
&= \frac{p^2 a_0}{p^2 a_0 + pqa_1 + q^2 a_2} = \frac{a_0}{a_0 + (q/p)a_1 + (q/p)^2 a_2} \\
&= \frac{a_0}{\sum_{i=0}^{2} a_i (1-r)^i} = \frac{a_0}{\sum_{i=0}^{2} a_i \bar{r}^i}.
\end{aligned}$$

This completes the proof for $n = 2$.

We now prove (1.5) for arbitrary $n \geq 2$. Let $P^* := P^*(\mathrm{R}_1, \ldots, \mathrm{R}_n)$. Applying Bayes Theorem, we obtain:

$$\boldsymbol{P}^* = \frac{\boldsymbol{P}(\mathrm{REPR}_1, \ldots, \mathrm{REPR}_n | \mathrm{R}_1, \ldots, \mathrm{R}_n)\ \boldsymbol{P}(\mathrm{R}_1, \ldots, \mathrm{R}_n)}{\boldsymbol{P}(\mathrm{REPR}_1, \ldots, \mathrm{REPR}_n)}.$$

We expand the denominator:

$$\boldsymbol{P}^* = \frac{\boldsymbol{P}(\mathrm{REPR}_1, \ldots, \mathrm{REPR}_n | \mathrm{R}_1, \ldots, \mathrm{R}_n)\ \boldsymbol{P}(\mathrm{R}_1, \ldots, \mathrm{R}_n)}{\sum_{R_1 = \mathrm{R}_1, \neg\mathrm{R}_1; \ldots; R_n = \mathrm{R}_n, \neg\mathrm{R}_n} \boldsymbol{P}(\mathrm{REPR}_1, \ldots, \mathrm{REPR}_n | R_1, \ldots, R_n)\ \boldsymbol{P}(R_1, \ldots, R_n)}.$$

The sum in the denominator ranges over *all* combinations of values of the propositional variables R_i (with $i = 1, \ldots, n$). Let us organize this sum as follows: We first group terms with the same number of negative values of propositional variables (and hence with the same number of positive values of propositional variables) together. We then use the following **lemma** (proven below) which generalizes (A.2) through (A.5) above:

$$\boldsymbol{P}(\mathrm{REPR}_1, \ldots, \mathrm{REPR}_n | R_1, \ldots, R_n) = \boldsymbol{P}(\mathrm{REPR}_1 | R_1) \cdot \ldots \cdot \boldsymbol{P}(\mathrm{REPR}_n | R_n). \quad \text{(A.6)}$$

Hence we obtain:

$$\boldsymbol{P}^* = \frac{\boldsymbol{P}(\mathrm{REPR}_1 | \mathrm{R}_1) \cdot \ldots \cdot \boldsymbol{P}(\mathrm{REPR}_n | \mathrm{R}_n)\ \boldsymbol{P}(\mathrm{R}_1, \ldots, \mathrm{R}_n)}{\sum_{i=0}^{n} \sum_{i\ \neg\mathrm{R}_k s,\ n-i\ \mathrm{R}_k s} \boldsymbol{P}(\mathrm{REPR}_1 | R_1) \cdot \ldots \cdot \boldsymbol{P}(\mathrm{REPR}_n | R_n)\ \boldsymbol{P}(R_1, \ldots, R_n)}.$$

The second sum in the denominator runs over all combinations of i propositional variables R_k having negative values, and $n - i$ propositional variables R_k having positive values. We divide the numerator and the denominator by the product $P(\mathrm{REPR}_1|\mathrm{R}_1)\cdot\cdots\cdot P(\mathrm{REPR}_n|\mathrm{R}_n) \neq 0$ and apply the definitions of the a_is and r:

$$P^* = \frac{a_0}{\sum_{i=0}^{n} (q/p)^i \sum_{i\,\neg\mathrm{R}_k s,\; n-i\,\mathrm{R}_k s} P(R_1, \ldots, R_n)}$$
$$= \frac{a_0}{\sum_{i=0}^{n} a_i (1-r)^i} = \frac{a_0}{\sum_{i=0}^{n} a_i \bar{r}^i}$$

The **lemma** (A.6) is proven by applying the chain rule of the probability calculus,

$$P(\mathrm{REPR}_1, \ldots, \mathrm{REPR}_n | R_1, \ldots, R_n) = P(\mathrm{REPR}_1 | \mathrm{REPR}_2, \ldots, \mathrm{REPR}_n, R_1, \ldots, R_n).$$
$$P(\mathrm{REPR}_2 | \mathrm{REPR}_3, \ldots, \mathrm{REPR}_n, R_1, \ldots, R_n)\cdot\cdots\cdot P(\mathrm{REPR}_n | R_1, \ldots, R_n),$$

and applying the independence assumptions in (1.3) to all factors:

$$P(\mathrm{REPR}_1, \ldots, \mathrm{REPR}_n | R_1, \ldots, R_n) = P(\mathrm{REPR}_1 | R_1 \ldots)\cdot\cdots\cdot P(\mathrm{REPR}_n | R_n).$$

This completes the proof of the general case.

B. PROOF OF CHAPTER 2

B.1. *Eq. (2.12)*

Let

$$c_r(\mathrm{S}) = \frac{a_0 + \bar{a}_0 \bar{r}^n}{\sum_{i=0}^{n} a_i \bar{r}^i}, \quad c_r(\mathrm{S}') = \frac{a'_0 + \bar{a}'_0 \bar{r}^n}{\sum_{i=0}^{n} a'_i \bar{r}^i}.$$

S is more coherent than or equally coherent as S′ if and only if

$$\Delta_0 := c_r(\mathrm{S}) - c_r(\mathrm{S}') \geq 0 \;\forall r \in (0, 1).$$

Since the denominators of $c_r(\mathrm{S})$ and $c_r(\mathrm{S}')$ are greater than 0 for $r \in (0, 1)$, it suffices for our purposes to study when

$$\Delta = \sum_{i=0}^{n} a_i \bar{r}^i \cdot \sum_{i=0}^{n} a'_i \bar{r}^i \cdot \Delta_0 \geq 0.$$

Let

$$D_i := a_0\, a'_i - a'_0\, a_i, \;\; \delta_i := a'_i - a_i. \tag{B.1}$$

Note that since $\sum_{i=0}^{n} a_i = \sum_{i=0}^{n} a'_i = 1$,

$$\sum_{i=0}^{n} D_i = a_0 - a'_0 = -\delta_0, \;\; \sum_{i=0}^{n} \delta_i = 0, \;\; D_0 = 0. \tag{B.2}$$

Using (B.1) and (B.2) one obtains:

$$\begin{aligned}
\Delta &= \sum_{i=0}^{n} (a_0\, a'_i\, \bar{r}^{\,i} + a'_i\, \bar{r}^{\,n+i} - a_0\, a'_i\, \bar{r}^{\,n+i} - a'_0\, a_i\, \bar{r}^{\,i} - a_i\, \bar{r}^{\,n+i} + a'_0\, a_i\, \bar{r}^{\,n+i}) \\
&= \sum_{i=0}^{n} [\delta_i \bar{r}^n + D_i(1 - \bar{r}^n)]\bar{r}^i \\
&= \delta_0 \bar{r}^n + \delta_n \bar{r}^{2n} + D_n(\bar{r}^n - \bar{r}^{2n}) + \sum_{i=1}^{n-1} [\delta_i \bar{r}^n + D_i(1 - \bar{r}^n)]\bar{r}^i.
\end{aligned}$$

Using the formulae in (B.2) we get $D_n = -\delta_0 - \sum_{i=1}^{n-1} D_i$ and $\delta_n = \delta_0 - \sum_{i=1}^{n-1} \delta_i$; so we obtain after some algebraic manipulations:

$$\Delta = \sum_{i=1}^{n-1} (\bar{r}^i - \bar{r}^n)[\delta_i \bar{r}^n + D_i(1 - \bar{r}^n)].$$

Since $\bar{r}^i - \bar{r}^n > 0$ for $i < n$ and $r \in (0, 1)$, a sufficient (and, for $n = 2$, also necessary) condition for S being more coherent than or equally coherent as S′ is

$$\delta_i \bar{r}^n + D_i(1 - \bar{r}^n) \geq 0 \;\; \forall i = 1, \ldots, n-1 \text{ and } \forall r \in (0, 1). \tag{B.3}$$

Let $\bar{r}^n =: \lambda \in (0, 1)$ and $f_i(\lambda) := \delta_i \lambda + D_i(1 - \lambda)$ $\forall i = 1, \ldots, n-1$. Since $f_i(\lambda) \geq 0$ and is monotonic over the range $(0, 1)$,

$$D_i = \lim_{\lambda \searrow 0} f_i(\lambda) \geq 0 \text{ and } \delta_i = \lim_{\lambda \nearrow 1} f_i(\lambda) \geq 0 \;\; \forall i = 1, \ldots, n-1.$$

Hence, (B.3) has the solution

$$D_i \geq 0 \wedge \delta_i \geq 0 \; \forall i = 1, \ldots, n-1.$$

We replace D_i and δ_i by the expressions given in (B.1) and obtain:

$$a_i'/a_i \geq a_0'/a_0 \wedge a_i'/a_i > 1 \quad \forall i = 1, \ldots, n-1.$$

Hence, S is more coherent than or equally coherent as S′ if (for $n = 2$: if and only if)

$$a_i'/a_i > \max(1, a_0'/a_0) \; \forall i = 1, \ldots, n-1. \qquad \text{(B.4)}$$

C. PROOFS OF CHAPTER 3

C.1. *Eq. (3.2)*

We proceed in two steps. First, we apply the definition of conditional probability to (3.1):

$$\frac{P(\text{HYP}, \neg\text{REL})}{P(\neg\text{REL})} = \frac{P(\text{HYP}, \neg\text{REL}, \text{REP})}{P(\neg\text{REL}, \text{REP})}.$$

Hence,

$$\frac{P(\neg\text{REL}, \text{REP})}{P(\neg\text{REL})} = \frac{P(\text{HYP}, \neg\text{REL}, \text{REP})}{P(\text{HYP}, \neg\text{REL})} = P(\text{REP}|\text{HYP}, \neg\text{REL}). \qquad \text{(C.1)}$$

Second, we apply the negation rule to (3.1):

$$P(\neg\text{HYP}|\neg\text{REL}) = P(\neg\text{HYP}|\text{REP}, \neg\text{REL}).$$

Using the definition of conditional probability, we obtain:

$$\frac{P(\neg\text{HYP}, \neg\text{REL})}{P(\neg\text{REL})} = \frac{P(\neg\text{HYP}, \neg\text{REL}, \text{REP})}{P(\neg\text{REL}, \text{REP})}.$$

Hence,

$$\frac{P(\neg \text{REL}, \text{REP})}{P(\neg \text{REL})} = \frac{P(\neg \text{HYP}, \neg \text{REL}, \text{REP})}{P(\neg \text{HYP}, \neg \text{REL})} = P(\text{REP}|\neg \text{HYP}, \neg \text{REL}). \tag{C.2}$$

The result obtains by comparing (C.1) and (C.2).

C.2. *Eq. (3.28)*

We apply Bayes Theorem and the independence assumptions in (3.26) and (3.27) (see also the Bayesian Network in Figure 3.6):

$$\begin{aligned} P'^{*(n)}(\text{REL}) &= \frac{P'(\text{REP}_1, \ldots, \text{REP}_n|\text{REL})\, P'(\text{REL})}{\sum_{REL} P'(\text{REP}_1, \ldots, \text{REP}_n|REL)\, P'(REL)} \\ &= \frac{\sum_{HYP} P'(\text{REP}_1, \ldots, \text{REP}_n|HYP, \text{REL})\, P'(\text{REL})\, P'(HYP)}{\sum_{HYP,\, REL} P'(\text{REP}_1, \ldots, \text{REP}_n|HYP, REL)\, P'(REL)\, P'(HYP)} \\ &= \frac{P'(\text{REL}) \sum_{HYP} P'(HYP) \prod_{i=1}^{n} P'(\text{REP}_i|HYP, \text{REL})}{\sum_{HYP,\, REL} P'(HYP)\, P'(REL) \prod_{i=1}^{n} P'(\text{REP}_i|HYP, REL)} \\ &= \frac{h\rho}{h\rho + a^n \bar{\rho}}. \end{aligned}$$

C.3. *Eq. (3.42)*

We use the conditional independences in the Bayesian Network in Figure 3.7,

$$P^{*(n-1)}(\text{REL}_n) = \frac{P(\text{REL}_1, \ldots, \text{REL}_n)}{P(\text{REL}_1, \ldots, \text{REL}_{n-1})},$$

and calculate:

$$\begin{aligned} &P(\text{REL}_1, \ldots, \text{REL}_n) \\ &= \sum_{HYP,\, SR} P(HYP)\, P(SR) \sum_{REP_1, \ldots, REP_n} \prod_{i=1}^{n} P(REP_i|HYP, \text{REL}_i)\, P(\text{REL}_i|SR) \\ &= \sum_{HYP,\, SR} P(HYP)\, P(SR) \prod_{i=1}^{n} \left[\left(\sum_{REP_i} P(REP_i|HYP, \text{REL}_i) \right) P(\text{REL}_i|SR) \right] \end{aligned}$$

$$= \sum_{SR} \prod_{i=1}^{n} P(\text{REL}_i|SR)\ P(SR)$$

$$= us^n + \bar{u}t^n. \tag{C.3}$$

Similarly, we obtain:

$$P(\text{REL}_1, \ldots, \text{REL}_{n-1}) = us^{n-1} + \bar{u}t^{n-1}. \tag{C.4}$$

From (C.3) and (C.4):

$$P^{*(n-1)}(\text{REL}_n) = \frac{us^n + \bar{u}t^n}{us^{n-1} + \bar{u}t^{n-1}}.$$

C.4. *Eq. (3.43)*

By the definition of conditional probability:

$$P^{*(n)}(\text{REL}_n) = \frac{P(\text{REL}_n, \text{REP}_1, \ldots, \text{REP}_n)}{P(\text{REP}_1, \ldots, \text{REP}_n)}.$$

We calculate the expression in the numerator and the denominator separately. We apply the conditional independences in the Bayesian Network in Figure 3.7. First,

$$P(\text{REL}_n, \text{REP}_1, \ldots, \text{REP}_n)$$

$$= \sum_{HYP,\, SR,\, REL_1, \ldots, REL_{n-1}} P(HYP, REL_1, \ldots, REL_{n-1}, \text{REL}_n, \text{REP}_1, \ldots, \text{REP}_n, SR)$$

$$= \sum_{HYP,\, SR,\, REL_1, \ldots, REL_{n-1}} P(HYP)\ P(SR)\ P(\text{REP}_n|HYP, \text{REL}_n)\ P(\text{REL}_n|SR)$$

$$\times \prod_{i=1}^{n-1} P(\text{REP}_i|HYP, REL_i) P(REL_i|SR)$$

$$= h[us(s + a\bar{s})^{n-1} + \bar{u}t(t + a\bar{t})^{n-1}]. \tag{C.5}$$

Secondly,

$$
\begin{aligned}
&\boldsymbol{P}(\mathrm{REP}_1, \ldots, \mathrm{REP}_\mathrm{n}) \\
&= \sum_{HYP,\, SR,\, REL_1, \ldots, REL_n} \boldsymbol{P}(HYP, REL_1, \ldots, REL_n, \mathrm{REP}_1, \ldots, \mathrm{REP}_\mathrm{n}, SR) \\
&= \sum_{HYP,\, SR,\, REL_1, \ldots, REL_n} \boldsymbol{P}(HYP)\boldsymbol{P}(SR) \prod_{i=1}^{n} \boldsymbol{P}(\mathrm{REP}_\mathrm{i}|HYP,\ REL_\mathrm{i})\boldsymbol{P}(REL_\mathrm{i}|SR) \\
&= h[u(s + a\bar{s})^n + \bar{u}(t + a\bar{t})^n] + \bar{h}[u(a\bar{s})^n + \bar{u}(a\bar{t})^n]. \qquad \text{(C.6)}
\end{aligned}
$$

Hence, from (C.5) and (C.6),

$$
\boldsymbol{P}^{*(n)}(\mathrm{REL_n}) = \frac{h[us(s + a\bar{s})^{n-1} + \bar{u}t(t + a\bar{t})^{n-1}]}{h[u(s + a\bar{s})^n + \bar{u}(t + a\bar{t})^n] + \bar{h}[u(a\bar{s})^n + \bar{u}(a\bar{t})^n]}.
$$

C.5. *Jury Voting*

We calculate the posterior probability of the hypothesis after getting n_+ positive reports and n_- negative reports. We apply Bayes Theorem:

$$
\boldsymbol{P}(\mathrm{HYP}|n_+\mathrm{REPs}, n_-\neg\mathrm{REPs}) = \frac{h}{h + \bar{h}x},
$$

with the likelihood ratio

$$
x = \frac{\boldsymbol{P}(n_+\mathrm{REPs}, n_-\neg\mathrm{REPs}|\neg\mathrm{HYP})}{\boldsymbol{P}(n_+\mathrm{REPs}, n_-\neg\mathrm{REPs}|\mathrm{HYP})}.
$$

By the conditional independences in the Bayesian Network in Figure 3.7:

$$
\begin{aligned}
&\boldsymbol{P}(n_+\mathrm{REPs}, n_-\neg\mathrm{REPs}|\mathrm{HYP}) \\
&= \binom{n}{n_+} \cdot \sum_{SR} \psi(\mathrm{HYP},\ \mathrm{REP}, SR)^{n_+} \psi(\mathrm{HYP}, \neg\mathrm{REP}, SR)^{n_-} \boldsymbol{P}(SR)
\end{aligned}
$$

with

$$
\psi(\mathrm{HYP}, \mathrm{REP}, SR) = \sum_{REL} \boldsymbol{P}(\mathrm{REP}|\mathrm{HYP},\ REL)\ \boldsymbol{P}(REL|SR)
$$

and an analogous expression for ψ(HYP, ¬REP, *SR*). Hence,

$$P(n_+\text{REPs}, n_-\neg\text{REPs}|\text{HYP}) = \binom{n}{n_+} \cdot [u(s + a\bar{s})^{n_+}(\bar{a}\bar{s})^{n_-} + \bar{u}(t + a\bar{t})^{n_+}(\bar{a}\bar{t})^{n_-}].$$

Similarly,

$$P(n_+\text{REPs}, n_-\neg\text{REPs}|\neg\text{HYP}) = \binom{n}{n_+} \cdot [u(a\bar{s})^{n_+}(s + \bar{a}\bar{s})^{n_-} + \bar{u}(a\bar{t})^{n_+}(t + \bar{a}\bar{t})^{n_-}].$$

Hence,

$$x = \frac{u(a\bar{s})^{n_+}(s + \bar{a}\bar{s})^{n_-} + \bar{u}(a\bar{t})^{n_+}(t + \bar{a}\bar{t})^{n_-}}{u(s + a\bar{s})^{n_+}(\bar{a}\bar{s})^{n_-} + \bar{u}(t + a\bar{t})^{n_+}(\bar{a}\bar{t})^{n_-}}.$$

For $a = 1/2$, we obtain:

$$x(a = 1/2) = \frac{u(1 - s)^{n_+}(1 + s)^{n_-} + \bar{u}(1 - t)^{n_+}(1 + t)^{n_-}}{u(1 + s)^{n_+}(1 - s)^{n_-} + \bar{u}(1 + t)^{n_+}(1 - t)^{n_-}}. \qquad \text{(C.7)}$$

We introduce the variables

$$n = n_+ + n_-, \quad \delta = n_+ - n_-,$$
$$s_\pm = \surd(1 \pm s), \quad t_\pm = \surd(1 \pm t),$$

and obtain after some algebraic manipulations:

$$x(a = 1/2) = \frac{u\left(\frac{s_+s_-}{t_+t_-}\right)^n\left(\frac{s_-}{s_+}\right)^\delta + \bar{u}\left(\frac{t_-}{t_+}\right)^\delta}{u\left(\frac{s_+s_-}{t_+t_-}\right)^n\left(\frac{s_+}{s_-}\right)^\delta + \bar{u}\left(\frac{t_+}{t_-}\right)^\delta}.$$

$x(a = 1/2)$ is a monotonically increasing function of n for positive δ and a decreasing function for negative δ. To study the limit $n \to \infty$, we observe that for $s > t$ it holds that $s_+s_- < t_+t_-$ and hence

$$\lim_{n\to\infty}\left(\frac{s_+s_-}{t_+t_-}\right)^n = 0,$$

and so

$$\lim_{n\to\infty} x(a=1/2) = \left(\frac{t_-}{t_+}\right)^{2\delta} = \left(\frac{1-t}{1+t}\right)^{\delta}.$$

We can stipulate that the witnesses are independent by setting $s = t$. If we insert this identity in (C.7), then we obtain the same likelihood ratio without specifying the limiting value:

$$x_{indep}(a=1/2) = \left(\frac{1-t}{1+t}\right)^{\delta}.$$

From (3.47) and (3.48) we know that for independent jurors,

$$x_{indep}(a=1/2) = \left(\frac{1-\rho}{1+\rho}\right)^{\delta}.$$

Hence, if there is positive relevance between the reliability levels of the jurors, then the likelihood ratio as $n \to \infty$ equals the likelihood ratio for independent voters with ρ set at $t = \mathbf{P}(\mathrm{REL_i}|\neg\mathrm{SR})$.

C.6. *Tversky and Kahneman's Linda*

We use the conditional independences in the Bayesian Network in Figure 3.13 and calculate:

$$\mathbf{P}(\mathrm{B}|\mathrm{REP_B}) = \frac{b(\rho + a\bar{\rho})}{b\rho + a\bar{\rho}}.$$

Similarly,

$$\mathbf{P}(\mathrm{B},\ \mathrm{F}|\mathrm{REP_B},\ \mathrm{REP_F}) = \frac{bf(\rho + \mathrm{a}^2\overline{\rho})}{bf\rho + a^2\overline{\rho}}.$$

To prove (3.50), we calculate the difference in (3.49) and obtain:

$$\Delta P = -\frac{ab[a^2\bar{f} + \bar{a}\,\bar{\rho}(a - f(a + \bar{b}))]}{(b\bar{\rho} + a\bar{\rho})(bf\,\bar{\rho} + a^2\bar{\rho})}.$$

Since $0 < a, b, f, r < 1$, (3.50) clearly holds.

D. PROOFS OF CHAPTER 4

To calculate the posterior probabilities and to determine the phase curves in this section, we first apply Bayes theorem to $\boldsymbol{P}^*(\mathrm{HYP}) = \boldsymbol{P}(\mathrm{HYP}|\mathrm{REP}_1, \mathrm{REP}_2)$:

$$\begin{aligned}\boldsymbol{P}^*(\mathrm{HYP}) &= \frac{\boldsymbol{P}(\mathrm{REP}_1, \mathrm{REP}_2|\mathrm{HYP})\ \boldsymbol{P}(\mathrm{HYP})}{\boldsymbol{P}(\mathrm{REP}_1, \mathrm{REP}_2|\mathrm{HYP})\ \boldsymbol{P}(\mathrm{HYP}) + \boldsymbol{P}(\mathrm{REP}_1,\ \mathrm{REP}_2|\neg\mathrm{HYP})\ \boldsymbol{P}(\neg\mathrm{HYP})} \\ &= \frac{h}{h + \bar{h}x},\end{aligned}$$

with the likelihood ratio

$$x = \frac{\boldsymbol{P}(\mathrm{REP}_1, \mathrm{REP}_2|\neg\mathrm{HYP})}{\boldsymbol{P}(\mathrm{REP}_1, \mathrm{REP}_2|\mathrm{HYP})}.$$

$\boldsymbol{P}^*(\mathrm{HYP})$ strictly monotonically increases from $\boldsymbol{P}^*(\mathrm{HYP}) = 0$ for $h = 0$ to $\boldsymbol{P}^*(\mathrm{HYP}) = 1$ for $h = 1$ if the likelihood ratio does not explicitly depend on h. In this case, $\boldsymbol{P}^*(\mathrm{HYP})$ is convex as a function of h for $0 < x < 1$ and linear for $x = 1$ and has slope $1/x$ at $h = 0$ and x at $h = 1$. To determine a phase curve, we calculate:

$$\frac{\boldsymbol{P}'^*(\mathrm{HYP}) - \boldsymbol{P}^*(\mathrm{HYP})}{\boldsymbol{P}'^*(\mathrm{HYP})\boldsymbol{P}^*(\mathrm{HYP})} = -(x' - x)\frac{\bar{h}}{h}. \qquad \text{(D.1)}$$

Hence, the phase curve for the posterior probabilities is completely determined by the phase curve for the likelihood ratios.

D.1. *Eq. (4.5)*

Following the strategy just outlined, the posterior probability is given by

$$\mathbf{P}^*(\text{HYP}) = \mathbf{P}(\text{HYP}|\text{REP}) = \frac{h}{h + \bar{h}x_0},$$

with the likelihood ratio

$$x_0 = \frac{\mathbf{P}(\text{REP}|\neg\text{HYP})}{\mathbf{P}(\text{REP}|\text{HYP})}.$$

We use the conditional independences in the Bayesian Network in Figure 4.1,

$$\begin{aligned}
\mathbf{P}(\text{REP}|\text{HYP}) &= \sum_{CON,\ REL} \mathbf{P}(CON|\text{HYP})\ \mathbf{P}(\text{REP}|CON,\ REL)\ \mathbf{P}(REL) \\
&= p(\rho + a\bar{\rho}) + \bar{p}a\bar{\rho} \\
&= p\rho + a\bar{\rho}.
\end{aligned}$$

Similarly,

$$\begin{aligned}
\mathbf{P}(\text{REP}|\neg\text{HYP}) &= \sum_{CON,\ REL} \mathbf{P}(CON|\neg\text{HYP})\ \mathbf{P}(\text{REP}|CON, REL)\ \mathbf{P}(REL) \\
&= q\rho + a\bar{\rho}.
\end{aligned}$$

Hence,

$$x_0 = \frac{q\rho + a\bar{\rho}}{p\rho + a\bar{\rho}}.$$

Note that $0 < x_0 < 1$ for $0 < q < p < 1$.

D.2. *Eq. (4.6)*

The expression for $\mathbf{P}^*(\text{HYP})$ can be written in a different form:

$$\mathbf{P}^*(\text{HYP}) = \frac{h(p\rho + a\bar{\rho})}{c\rho + a\bar{\rho}},$$

with $c := \mathbf{P}(\text{CON}) = hp + \bar{h}q$. We differentiate $\mathbf{P}^*(\text{HYP})$ twice with respect to ρ:

$$\frac{\partial}{\partial \rho} P^*(\text{HYP}) = \frac{ah\bar{h}(p-q)}{(c\rho + a\bar{\rho})^2},$$
$$\frac{\partial^2}{\partial \rho^2} P^*(\text{HYP}) = \frac{2ah\bar{h}(p-q)(a-c)}{(c\rho + a\bar{\rho})^3}.$$

Hence, $\mathbf{P}^*(\text{HYP})$ is a monotonically increasing function of ρ for $p > q$. It is concave for $a < c$, linear for $a = c$, and convex for $a > c$ for $p > q$.

D.3. *Eq. (4.9)*

We use the conditional independences in the Bayesian Network in Figure 4.4 to determine the likelihood ratio:

$$\begin{aligned}
&\mathbf{P}(\text{REP}_1, \text{REP}_2|\text{HYP}) \\
&= \sum_{CON,\ REL} \mathbf{P}(\text{HYP}|CON)\ \mathbf{P}(\text{REP}_1|CON,\ REL)\ \mathbf{P}(\text{REP}_2|CON, REL)\ \mathbf{P}(REL) \\
&= p(\rho + a^2\bar{\rho}) + \bar{p}a^2\bar{\rho} \\
&= p\rho + a^2\bar{\rho}.
\end{aligned}$$

Similarly,

$$\mathbf{P}(\text{REP}_1,\ \text{REP}_2|\neg\text{HYP}) = q\rho + a^2\bar{\rho}.$$

Hence,

$$x = \frac{q\rho + a^2\bar{\rho}}{p\rho + a^2\bar{\rho}}.$$

To prove (4.9), we calculate:

$$x - x_0 = -\frac{a\bar{a}(p-q)\rho\bar{\rho}}{(p\rho + a\bar{\rho})(p\rho + a^2\bar{\rho})}.$$

Since $0 < a,\ \rho < 1$ and $p > q$, (4.9) clearly holds by (D.1).

D.4. *Eq. (4.11)*

We use the conditional independences in the Bayesian Network in Figure 4.4 to determine the likelihood ratio:

$$\begin{aligned}
&\mathbf{P}'(\mathrm{REP}_1,\ \mathrm{REP}_2|\mathrm{HYP}) \\
&= \sum_{CON,\ REL_1,\ REL_2} \mathbf{P}'(\mathrm{HYP}|CON) \prod_{i=1}^{2} \mathbf{P}'(\mathrm{REP_i}|CON, REL_i)\ \mathbf{P}'(REL_i)\,. \\
&= p(\rho + a\overline{\rho})^2 + \overline{p}(a\overline{\rho})^2
\end{aligned}$$

Similarly,

$$\mathbf{P}'(\mathrm{REP}_1,\ \mathrm{REP}_2|\neg\mathrm{HYP}) = q(\rho + a\overline{\rho})^2 + \overline{q}(a\overline{\rho})^2.$$

Hence,

$$x' = \frac{q(\rho + a\overline{\rho})^2 + \overline{q}(a\overline{\rho})^2}{p(\rho + a\overline{\rho})^2 + \overline{p}(a\overline{\rho})^2}.$$

To prove (4.10), we calculate:

$$x' - x_0 = -\frac{a(a + \overline{a}\rho)\,(p - q)\rho\overline{\rho}}{(p\rho + a\overline{\rho})\,(p\,(\rho + a\overline{\rho})^2 + \overline{p}\,(a\overline{\rho})^2)}.$$

Since $0 < a,\ \rho < 1$ and $p > q$, (4.10) clearly holds by (D.1).

D.5. *Eq. (4.11)*

With the results of the last two appendices, we can calculate the difference:

$$x' - x = \frac{a^2(p - q)\rho\overline{\rho}[1 - 2\overline{a}\,\overline{\rho}\,]}{(p\rho + a^2\overline{\rho})(p(\rho + a\overline{\rho})^2 + \overline{p}(a\overline{\rho})^2)}.$$

Since $0 < a, \rho < 1$ and $p > q$, $\mathbf{P}'^*(\mathrm{HYP}) - \mathbf{P}^*(\mathrm{HYP}) > 0$ if and only if $1 - 2\overline{a}\,\overline{\rho} > 0$ by (D.1).

D.6. *Eq. (4.12)*

We use the conditional independences in the Bayesian Network in Figure 4.7 to determine the likelihood ratio:

$$\begin{aligned}
&\mathbf{P}'(\mathrm{REP}_1, \mathrm{REP}_2|\mathrm{HYP}) \\
&= \sum_{CON_1,\, REL_1} \sum_{CON_2,\, REL_2} \prod_{i=1}^{2} \mathbf{P}'(\mathrm{HYP}|CON_i)\ \mathbf{P}'(\mathrm{REP}_i|CON_i,\, REL_i)\ \mathbf{P}'(REL_i) \\
&= (p(\rho + a\overline{\rho}) + \overline{p}a\overline{\rho})^2 \\
&= (p\rho + a\overline{\rho})^2.
\end{aligned}$$

Similarly,

$$\mathbf{P}'(\mathrm{REP}_1, \mathrm{REP}_2|\neg\mathrm{HYP}) = (q\rho + a\overline{\rho})^2.$$

Hence,

$$x' = \left(\frac{q\rho + a\overline{\rho}}{p\rho + a\overline{\rho}}\right)^2 = x_0^2.$$

To prove (4.13), we calculate:

$$x' - x_0 = x_0^2 - x_0 = -x_0\overline{x}_0.$$

Since $p > q$, (4.13) clearly holds by (D.1).

D.7. *Eq. (4.14)*

We use the conditional independences in the Bayesian Network in Figure 4.6 to determine the likelihood ratio:

$$\begin{aligned}
&\mathbf{P}(\mathrm{REP}_1, \mathrm{REP}_2|\mathrm{HYP}) \\
&= \sum_{CON_1,\, CON_2,\, REL} \prod_{i=1}^{2} \mathbf{P}(\mathrm{HYP}|CON_i)\ \mathbf{P}(\mathrm{REP}_i|CON_i,\, REL)\ \mathbf{P}(REL) \\
&= p^2\rho + (ap + a\overline{p})^2\overline{\rho} \\
&= p^2\rho + a^2\overline{\rho}.
\end{aligned}$$

Similarly,

$$\boldsymbol{P}(\text{REP}_1, \text{REP}_2|\neg\text{HYP}) = q^2\rho + a^2\overline{\rho}.$$

Hence,

$$x' = \frac{q^2\rho + a^2\overline{\rho}}{p^2\rho + a^2\overline{\rho}}.$$

To prove (4.14), we calculate:

$$x - x_0 = -\frac{(p-q)\rho[pq\rho + a\overline{\rho}(p+q-a)]}{(p\rho + a\overline{\rho})(p^2\rho + a^2\overline{\rho})}.$$

Since $0 < a, \rho < 1$ and $p > q$, $\boldsymbol{\Delta P} > 0$ if and only if $pqr + a\overline{\rho}\ (p+q-a) > 0$ by (D.1). Note that $p + q > a$ is a sufficient condition for $\boldsymbol{\Delta P} > 0$.

D.8. *Eq. (4.14)*

With the results of the last two appendices, we can calculate the difference

$$x' - x = -\frac{a(p-q)\rho\overline{\rho}[(2a-p-q)a - 2(a-p)(a-q)\rho]}{(p^2\rho + a^2\overline{\rho})(p\rho + a\overline{\rho})^2}.$$

Since $0 < a, h, \rho < 1$ and $p > q$, $\boldsymbol{\Delta P} > 0$ if and only if $(2a-p-q)a - 2(a-p)(a-q)\rho > 0$ by (D.1).

D.9. *Eq. (4.15)*

The likelihood ratio for the Bayesian Network in Figure 4.12 can be obtained by replacing ρ by $\boldsymbol{P}(\text{AUX}) = ht_h + \overline{h}t_{\overline{h}}$ in the corresponding expression of our basic model:

$$x = \frac{a(1 - ht_h - \overline{h}t_{\overline{h}})}{ht_h + \overline{h}t_{\overline{h}} + a(1 - ht_h - \overline{h}t_{\overline{h}})}.$$

Let us now turn to the Bayesian Network in Figure 4.13. We calculate:

$$
\begin{aligned}
&\mathbf{P}(\text{REP}|\text{HYP})\\
&= \sum_{AUX,\,CON,\,REL} \mathbf{P}(REL|AUX)\ \mathbf{P}(AUX|\text{HYP})\ \mathbf{P}(\text{REP}|CON,\ REL)\ \mathbf{P}(CON|\text{HYP})\\
&= \sum_{REL} \mathbf{P}(\text{REP}|\text{CON},\ REL)\left[\sum_{AUX} \mathbf{P}(REL|AUX)\mathbf{P}(AUX|\text{HYP})\right]\\
&= \sum_{REL} \mathbf{P}(\text{REP}|\text{CON},\ REL)\ \mathbf{P}(REL|\text{HYP})\\
&= t_h + a\bar{t}_h
\end{aligned}
$$

Similarly,

$$\mathbf{P}(\text{REP}|\neg\text{HYP}) = a t_{\bar{h}}.$$

Here we made use of the identities

$$\mathbf{P}(\text{REP}|\text{HYP}) = t_h, \quad \mathbf{P}(\text{REP}|\neg\text{HYP}) = t_{\bar{h}}.$$

Hence,

$$x' = \frac{a t_{\bar{h}}}{t_h + a\bar{t}_h}.$$

We now construct the difference

$$x' - x = -\frac{a\bar{h}(t_h - t_{\bar{h}})[h - \bar{a}(1 - ht_h - \bar{h}t_{\bar{h}})]}{(ht_h + \bar{h}t_{\bar{h}} + a(1 - ht_h - \bar{h}t_{\bar{h}}))(t_h + a\bar{t}_h)}.$$

Since $0 < a, h < 1$ and $t_h > t_{\bar{h}}$, $\Delta \mathbf{P} > 0$ if and only if $h - \bar{a}\,(1 - ht_h - \bar{h}t_{\bar{h}}) > 0$ by (D.1).

E. PROOFS OF CHAPTER 5

E.1. *Eq. (5.1)*

We apply Bayes Theorem:

$$\mathbf{P}^{*(n)}(\text{HYP}) = \frac{\mathbf{P}(\text{HYP})}{\mathbf{P}(\text{HYP}) + \mathbf{P}(\neg\text{HYP})x},$$

with the likelihood ratio

$$x = \frac{P(\text{REP}_1, \ldots, \text{REP}_n | \neg\text{HYP})}{P(\text{REP}_1, \ldots, \text{REP}_n | \text{HYP})}.$$

We use the conditional independences in the Bayesian Network in Figure 3.5 to determine the likelihood ratio:

$$x = \frac{\sum_{REL_1, \ldots, REL_n} P(\text{REP}_1, \ldots, \text{REP}_n | \neg\text{HYP}, REL_1, \ldots, REL_n)\, P(REL_1, \ldots, REL_n)}{\sum_{REL_1, \ldots, REL_n} P(\text{REP}_1, \ldots, \text{REP}_n | \text{HYP}, REL_1, \ldots, REL_n)\, P(REL_1, \ldots, REL_n)}$$

$$= \frac{\prod_{i=1}^{n} \sum_{REL_i} P(\text{REP}_i | \neg\text{HYP}, REL_i)\, P(REL_i)}{\prod_{i=1}^{n} \sum_{REL_i} P(\text{REP}_i | \text{HYP}, REL_i)\, P(REL_i)} = \frac{\prod_{i=1}^{n} (1 - \rho_i) a}{\prod_{i=1}^{n} (\rho_i + a(1 - \rho_i))}$$

$$= \prod_{i=1}^{n} x_i,$$

with the likelihood ratios of the individual reports

$$x_i = \frac{a\bar{\rho}_i}{\rho_i + a\bar{\rho}_i}.$$

We obtain:

$$P^{*(n)}(\text{HYP}) = \frac{h}{h + \bar{h} \prod_{i=1}^{n} x_i}.$$

Note that (3.19) follows if all likelihood ratios are equal.

E.2. *Eq. (5.2)*

Without loss of generality, we present the proof for $P^{*(n)}(\text{REL}_1)$:

$$P^{*(n)}(\text{REL}_1) = \frac{P(\text{REL}_1, \text{REP}_1, \ldots, \text{REP}_n)}{P(\text{REP}_1, \ldots, \text{REP}_n)}$$

$$= \frac{\sum_{HYP, REL_2, \ldots, REL_n} P(HYP, \text{REL}_1, REL_2, \ldots, REL_n, \text{REP}_1, \ldots, \text{REP}_n)}{\sum_{HYP, REL_1, \ldots, REL_n} P(HYP, REL_1, \ldots, REL_n, \text{REP}_1, \ldots, \text{REP}_n)}$$

We calculate the expression in the numerator and the denominator separately, using the conditional independences in the Bayesian Network in Figure 3.5. First,

$$\boldsymbol{P}(\mathrm{REL}_1, \mathrm{REP}_1, \ldots, \mathrm{REP}_n) = \sum_{HYP} \boldsymbol{P}(HYP)\ \boldsymbol{P}(\mathrm{REP}_1|HYP, \mathrm{REL}_1)\ \boldsymbol{P}(\mathrm{REL}_1)$$

$$\times \sum_{REL_2, \ldots, REL_n} \prod_{i=2}^{n} \boldsymbol{P}(\mathrm{REP}_i|HYP, REL_i)\ \boldsymbol{P}(REL_i)$$

$$= h\rho_1 \prod_{i=2}^{n} (\rho_i + a\bar{\rho}_i). \tag{E.1}$$

Secondly,

$$\boldsymbol{P}(\mathrm{REP}_1, \ldots, \mathrm{REP}_n) = \sum_{HYP} \boldsymbol{P}(HYP)$$

$$\times \sum_{REL_1, \ldots, REL_n} \prod_{i=1}^{n} \boldsymbol{P}(\mathrm{REP}_i|\mathrm{HYP},\ REL_i)\ \boldsymbol{P}(REL_i)$$

$$= h \prod_{i=1}^{n} (\rho_i + a\bar{\rho}_i) + \bar{h} \prod_{i=1}^{n} (a\bar{\rho}_i). \tag{E.2}$$

From (E.1) and (E.2):

$$\begin{aligned}
\boldsymbol{P}^{*(n)}(\mathrm{REL}_1) &= \frac{h\rho_1 \prod_{i=2}^{n} (\rho_i + a\bar{\rho}_i)}{h \prod_{i=1}^{n} (\rho_i + a\bar{\rho}_i) + \bar{h} \prod_{i=1}^{n} (a\bar{\rho}_i)} \\
&= \frac{\rho_1}{\rho_1 + a\bar{\rho}_1} \cdot \frac{h \prod_{i=1}^{n} (\rho_i + a\bar{\rho}_i)}{h \prod_{i=1}^{n} (\rho_i + a\bar{\rho}_i) + \bar{h} \prod_{i=1}^{n} (a\bar{\rho}_i)} \\
&= \frac{h\bar{x}_1}{h + \bar{h} \prod_{i=1}^{n} x_i},
\end{aligned}$$

with the likelihood ratios

$$x_i = \frac{a\bar{\rho}_i}{\rho_i + a\bar{\rho}_i}.$$

Note that (3.16) follows if all likelihood ratios are equal.

E.3. *Eq. (5.3)*

The derivatives are:

$$\frac{\partial}{\partial a}\mathbf{P}*^{(n)}(\text{HYP}) = -\frac{1}{a\bar{h}}\mathbf{P}*^{(n)}(\text{HYP})\overline{\mathbf{P}*^{(n)}(\text{HYP})}\sum_{j=1}^{n}\bar{x}_j < 0,$$

$$\frac{\partial}{\partial a}\mathbf{P}*^{(n)}(\text{REL}_i) = -\frac{1}{a}\mathbf{P}*^{(n)}(\text{REL}_i)\left[x_i + \frac{1}{h}\overline{\mathbf{P}*^{(n)}(\text{HYP})}\sum_{j=1}^{n}x_j\right] < 0.$$

E.4. *Reliability Bound for n = 2*

We express the posterior probability as a function of the number m of suspects for two witnesses. Since $h = 1/m$, we obtain from (5.4):

$$\mathbf{P}_m*^{(2)}(\text{HYP}) = \left[1 + (m-1)\left(\frac{1-r}{(m-1)r+1}\right)^2\right]^{-1}.$$

$\mathbf{P}^{*(2)}_{m+1}(\text{HYP}) > \mathbf{P}^{*(2)}_{m}(\text{HYP})$ for $m \geq 2$ if and only if

$$(m-1)\left(\frac{1-r}{(m-1)r+1}\right)^2 > m\left(\frac{1-r}{mr+1}\right)^2 \quad \text{for } m \geq 2.$$

This inequality can be simplified to

$$r > \frac{1}{\sqrt{m(m-1)}} \quad \text{for } m \geq 2.$$

Hence, $\mathbf{P}^{*(2)}_{m+1}(\text{HYP}) > \mathbf{P}^{*(2)}_{m}(\text{HYP})$ for $m \geq 2$ if $r > 1/\sqrt{2} \approx .707$.

E.5. *Eq. (5.10)*

If all $\rho_i =: \rho$ (for $i = 1, \ldots, n$), the posterior probability of the hypothesis is:

$$\mathbf{P}*^{(n)}(\text{HYP}) = \frac{h}{h + \bar{h}x_0^n},$$

with the likelihood ratio

$$x_0 = \frac{h\overline{\rho}}{\rho + h\overline{\rho}}.$$

Differentiating $\boldsymbol{P}^{*(n)}(\text{HYP})$ with respect to h, one obtains:

$$\frac{\partial}{\partial h}\boldsymbol{P}^{*(n)}(\text{HYP}) = \frac{x_0^n(1 - n\bar{h}\overline{x}_0)}{\left(h + \bar{h}x_0^n\right)^2}.$$

We set this partial derivative to 0 and solve for $h \in (0, 1)$:

$$h_{\min} = \frac{(n-1)\rho}{1 + (n-1)\rho}.$$

By examining the second derivative of $\boldsymbol{P}^{*(n)}(\text{HYP})$ with respect to h, it can be shown that this value of h corresponds to the minimum of $\boldsymbol{P}^{*(n)}$ (HYP).

E.6. *Numerical Details*

In order to plot Figure 5.5, we used the following observation: For $h > .5$, the expected value of $\boldsymbol{P}^{*(n)}(\text{HYP})$ for n independent witnesses is

$$\left\langle \boldsymbol{P}^{*(n)}(\text{HYP})\right\rangle = \sum_{k=0}^{n}(-1)^k\left(\frac{\bar{h}}{h}\right)^k I_k^n(h),$$

with

$$I_k(h) = \int_0^1 \frac{hx^k dx}{(h + \bar{h}x)^2}.$$

Proof:
Introducing the likelihood ratios as new integration variables,

$$x_i = \frac{h\overline{\rho_i}}{\rho_i + h\overline{\rho_i}},$$

one obtains,

$$\begin{aligned}\left\langle \mathbf{P}^{*(n)}(\mathrm{HYP})\right\rangle &= \int_0^1 d\rho_1 \ldots \int_0^1 d\rho_n \mathbf{P}^{*(n)}(\mathrm{HYP}) \\ &= \int_0^1 \ldots \int_0^1 \frac{1}{1+\frac{\bar{h}}{h}\prod_{i=1}^n x_i} \cdot \prod_{j=1}^n \frac{h dx_j}{(h+\bar{h}x_j)^2}.\end{aligned} \tag{E.3}$$

Since $0 < x_i < 1$, the expression $\frac{\bar{h}}{h}\prod_{i=1}^n x_i < 1$, if and only if $h > .5$. Hence, the first expression under the integral can be expanded in an infinite sum:

$$\left(1+\frac{\bar{h}}{h}\prod_{i=1}^n x_i\right)^{-1} = \sum_{k=0}^n (-1)^k \left(\frac{\bar{h}}{h}\right)^k \prod_{i=1}^n x_i^k, \tag{E.4}$$

for $h > .5$.

Substituting (E.4) in (E.3), we obtain:

$$\begin{aligned}\left\langle \mathbf{P}^{*(n)}(\mathrm{HYP})\right\rangle &= \sum_{k=0}^n (-1)^k \left(\frac{\bar{h}}{h}\right)^k \prod_{i=1}^n \int_0^1 \frac{h x_i^k dx_i}{(h+\bar{h}x_i)^2} \\ &= \sum_{k=0}^n (-1)^k \left(\frac{\bar{h}}{h}\right)^k \left[\int_0^1 \frac{h x^k dx}{(h+\bar{h}x)^2}\right]^n \\ &= \sum_{k=0}^n (-1)^k \left(\frac{\bar{h}}{h}\right)^k I_k^n(h).\end{aligned}$$

To determine the minimum of $\left\langle \mathbf{P}^{*(n)}(\mathrm{HYP})\right\rangle$, we did the exact n-dimensional numerical integration for $n = 2, 3, 4$, and 5 using the software *Mathematica* and then applied the built-in function **FindMinimum**. Following this procedure, the computation time grows exponentially with n. For $n > 5$, the minimum turns out to be at values of $h > .5$, and the above expression for $\left\langle \mathbf{P}^{*(n)}(\mathrm{HYP})\right\rangle$ can be used. The integrals $I_k(h)$ can easily be computed recursively:

$$
\begin{aligned}
I_0(h) &= 1, \\
I_1(h) &= \frac{h}{\bar{h}^2} \cdot [\ln(1/h) - \bar{h}], \\
I_{k+1}(h) &= \frac{h}{\bar{h}k} \cdot [1 - (k+1) I_k(h)] \quad \text{for } k \geq 1.
\end{aligned}
$$

Note that the computation time grows now only linearly with n.

References

Bernardo, J., and Smith, A. (2001). *Bayesian Theory.* Weinheim: Wiley.

Bonjour, L. (1985). *The Structure of Empirical Knowledge.* Cambridge, Mass.: Harvard University Press.

Bovens, L., Fitelson, B., Hartmann, S., and Snyder, J. (2002). 'Too Odd (Not) to Be True: A Reply to Olsson', *British Journal for the Philosophy of Science*, 53: 539–63.

—— and Hartmann, S. (2001). 'A Probabilistic Theory of the Coherence of an Information Set', in A. Beckermann and C. Nimtz (*eds.*), *Argument and Analysis—A Selection of Papers contributed to the Sections of the 4th International Congress of the Society for Analytical Philosophy,* 195–206. http://www.gap-im-netz.de/gap4Konf/Proceedings4/Proc.htm.

———— (2002). 'Bayesian Networks and the Problem of Unreliable Instruments', *Philosophy of Science*, 69: 29–72.

———— (2003). 'Solving the Riddle of Coherence', *Mind*, 112: 601–34.

—— and Olsson, E. J. (2000). 'Coherentism, Reliability and Bayesian Networks', *Mind*, 109: 685–719.

———— (2002). 'Believing More, Risking Less: On Coherence, Truth and Non-Trivial Extensions', *Erkenntnis*, 57: 137–50.

Cohen, L. J. (1977). *The Probable and the Provable.* Oxford: Clarendon.

Dawid, A. P. (1979). 'Conditional Independence in Statistical Theory', *Journal of the Royal Statistical Society,* ser. B 41 (no. 1): 1–31.

Dorling, J. (1996). 'Further Illustrations of the Bayesian Solution of Duhem's Problem.' http://www.princeton.edu/~bayesway/Dorling/dorling.html.

Earman, J. (1992). *Bayes or Bust? A Critical Examination of Bayesian Confirmation Theory.* Cambridge, Mass.: MIT Press.

—— (2000). *Hume's Abject Failure.* Oxford: Oxford University Press.

Edman, M. (1973). 'Adding Independent Pieces of Evidence', in S. Halldén (ed.), *Modality, Morality and Other Problems of Sense and Nonsense.* Lund: Gleerup, 180–8.

Eells, E., and Fitelson, B. (2001). 'Symmetries and Asymmetries in Evidential Support', *Philosophical Studies* 107: 129–42.

Ekelöf, P. O. (1983). 'My Thoughts on Evidentiary Value', in P. Gärdenfors, B. Hansson, and N. Sahlin (eds.), *Evidentiary Value: Philosophical, Judicial and Psychological Aspects of a Theory—Essays Dedicated to Sören Halldén on his Sixtieth Birthday.* Lund: Gleerup, 9–26.

Fitelson, B. (1996). 'Wayne, Horwich and Evidential Diversity', *Philosophy of Science*, 63: 652–60.

——(1999). 'The Plurality of Bayesian Measures of Confirmation and the Problem of Measure Sensitivity', *Philosophy of Science*, 63: 652–60.

——(2001). *'Studies in Bayesian Confirmation Theory'*, Ph.D. Dissertation in Philosophy, Madison, Wis.: University of Wisconsin.

——(2003). 'A Probabilistic Theory of Coherence', *Analysis*, 63: 194–99.

Gillies, D. (2000). *Philosophical Theories of Probability.* London: Routledge.

Hansson, B. (1983). 'Epistemology and Evidence', in P. Gärdenfors, B. Hansson, and N. Sahlin (eds.), *Evidentiary Value: Philosophical, Judicial and Psychological Aspects of a Theory—Essays Dedicated to Sören Halldén on his Sixtieth Birthday.* Lund: Gleerup—Library of *Theoria*, 15: 75–97.

Hawthorne, J. (1996). 'Voting in Search of the Public Good: The Probabilistic Logic of Majority Judgments', unpublished manuscript.

Hilden, J. (2002). 'Improbable Agreement between Witnesses: Chasing Down Statistical Paradoxes', unpublished manuscript.

Horwich, P. (1982). *Probability and Evidence.* Princeton: Princeton University Press.

Howson, C., and Urbach, P. ([1989] 1993). *Scientific Reasoning – The Bayesian Approach*, 2nd ed. Chicago: Open Court.

Jensen, F. V. (1996). *An Introduction to Bayesian Networks.* Berlin: Springer.

Kuhn, T. S. (1977). 'Objectivity, Value Judgment, and Theory Choice', in T. S. Kuhn, *The Essential Tension: Selected Studies in Scientific Tradition and Change.* Chicago: University of Chicago Press, 320–9.

Lewis, C. I. (1946). *An Analysis of Knowledge and Valuation.* LaSalle, Ill.: Open Court.

List, C. (2003). 'On the Significance of the Absolute Margin', *British Journal for the Philosophy of Science*, forthcoming.

Neapolitan, R. E. (1990). *Probabilistic Reasoning in Expert Systems.* New York: Wiley.

Olsson, E. J. (2002*a*). 'Corroborating Testimony, Probability and Surprise', *British Journal for the Philosophy of Science*, 53: 273–88.

——(2002*b*). 'What is the Problem of Coherence and Truth?' *Journal of Philosophy*, 94: 246–72.

——(2002*c*). 'Corroborating Testimony and Ignorance: A Reply to Bovens, Fitelson, Hartmann and Snyder', *British Journal for the Philosophy of Science*, 53: 565–72.

Pearl, J. ([1988] 1997) *Probabilistic Reasoning in Intelligent Systems: Networks of Plausible Inference*, 2nd ed. San Francisco: Morgan Kaufmann.

——(2000). *Causality: Models, Reasoning, and Inference.* Cambridge: Cambridge University Press.

Quine, W. V. O. (1960). *Word and Object.* Cambridge, Mass.: MIT Press.

Reichenbach, H. (1956). *The Direction of Time*. Berkeley: University of California Press.

Robert, C. P. (2001 [1994]). *The Bayesian Choice–From Decision-Theoretic Foundations to Computational Implementation*, 2nd ed. New York: Springer.

Salmon, W. C. (1998). 'Probabilistic Causality', in W. Salmon, *Causality and Explanation*. New York: Oxford University Press, ch. 14.

——(1990). 'Rationality and Objectivity in Science *or* Tom Kuhn Meets Tom Bayes', in C. W. Savage (ed.), *Scientific Theories*. Minneapolis: University of Minnesota Press, 175–204.

Schum, D. A. (1988). 'Probability and the Processes of Discovery, Proof and Choice', in P. Tillers and E. D. Green (eds.), *Probability and Inference in the Law of Evidence – The Uses and Limits of Bayesianism*. London: Kluwer, 213–70.

Sen, A. (1970). *Collective Choice and Social Welfare*. San Francisco: Holden-Day.

Shogenji, T. (1999). 'Is Coherence Truth-Conducive?' *Analysis*, 59: 338–45.

Spirtes, P., Glymour, C., and Scheines, R. (2000). *Causation, Prediction, and Search*, 2nd ed. Cambridge, Mass.: MIT Press.

Spohn, W. (1980). 'Stochastic Independence, Causal Independence, and Shieldability', *Journal of Philosophical Logic*, 9: 73–99.

Tversky, A., and Kahneman, D. (2002). 'Extensional versus Intuitive Reasoning: The Conjunction Fallacy', in T. Gilovich, D. Griffin, and D. Kahneman (eds.), *Heuristic and Biases – The Psychology of Intuitive Judgment*. Cambridge: Cambridge University Press, 19–48 edited version of a paper that originally appeared in *Psychological Review*, 90 (1983): 293–315.

Wayne, A. (1995). 'Bayesianism and Diverse Evidence', *Philosophy of Science*, 62: 111–21.

Index